Mark Palmer

THE FIRST BRAIN

The First Brain

THE NEUROSCIENCE OF PLANARIANS

Oné R. Pagán

UNIVERSITY PRESS

Oxford University Press is a department of the University of
Oxford. It furthers the University's objective of excellence in research,
scholarship, and education by publishing worldwide.

Oxford New York
Auckland Cape Town Dar es Salaam Hong Kong Karachi
Kuala Lumpur Madrid Melbourne Mexico City Nairobi
New Delhi Shanghai Taipei Toronto

With offices in
Argentina Austria Brazil Chile Czech Republic France Greece
Guatemala Hungary Italy Japan Poland Portugal Singapore
South Korea Switzerland Thailand Turkey Ukraine Vietnam

Oxford is a registered trademark of Oxford University Press
in the UK and certain other countries.

Published in the United States of America by
Oxford University Press
198 Madison Avenue, New York, NY 10016

© Oxford University Press 2014

All rights reserved. No part of this publication may be reproduced, stored in
a retrieval system, or transmitted, in any form or by any means, without the prior
permission in writing of Oxford University Press, or as expressly permitted by law,
by license, or under terms agreed with the appropriate reproduction rights organization.
Inquiries concerning reproduction outside the scope of the above should be sent to the
Rights Department, Oxford University Press, at the address above.

You must not circulate this work in any other form
and you must impose this same condition on any acquirer.

Library of Congress Cataloging-in-Publication Data
Pagán, Oné R.
The first brain : the neuroscience of planarians / Oné R. Pagán.
   pages cm
Includes bibliographical references and index.
ISBN 978–0–19–996504–5 (alk. paper)
1. Brain—Evolution. 2. Developmental neurophysiology. 3. Turbellaria. I. Title.
QP376.P29 2014
612.8′2—dc23
2013031621

9 8 7 6 5 4 3 2 1
Printed in the United States of America
on acid-free paper

To my late father, Mr. Onésimo Pagán, who would have bragged endlessly about this book, and to my youngest son, Andrew (Andy), who, upon learning that I was writing my first book, immediately became its biggest fan.

Also, to my son, Reynaldo, who gave me the brightest smile when I told him about it, and to my daughter, Giselle Vanessa, whose very first words when hearing the news were: "Dad, you're going to be famous!"

Contents

Foreword by Robert B. Raffa, PhD xi
Foreword by Harvey B. Sarnat, MD xiii
Acknowledgments xvii

Introduction xxi

PART I | FUNDAMENTALS

1. *Science* 3
    Science Does Not Exist 3
    Reductionism and Its Advantages and Limits 6
    On Theories and Theories 12
    Evolution and Life 14

2. *Biomedical Research* 19
    Biochemistry and Cell Biology Lite 19
    How We Classify Life 24
    Animal Models and Biomedical Research 26
    Basic or Fundamental Research? 33

PART II | THE SCIENCE OF THE BRAIN

3. *An Introduction to the Neurosciences* 41
    Neuroscience or Neurobiology? 41
    Neurons 42
    The Spaniard and the Italian 44
    Excitable Cells and Electrophysiology 50
    Synapses and Chemical Neurotransmission 53

Is a Nervous System Absolutely Necessary for Survival? 57
Plant Neurobiology 59

4. *The Human Brain* 62
What Exactly Is a Brain? 62
The Human Brain and Nervous System 64
A Brief History of What People Thought of Their Brains 66
The Complexity of the Human Brain/On Really BIG Numbers 71

5. *Some Brief Thoughts on Pharmacology* 78
What Is Pharmacology? 78
Psychopharmacology, Plant Style 82
Animal Models in Pharmacology 90

PART III | PLANARIANS

6. *Planarians* 95
What Is a Flatworm? 95
Flatworm Fossil Records 101
What Is a Planarian? 102
Early Works with Planarians (1700s–1800s) 108
Planariologists: Three "Personal" Connections 111

7. *Planarians in Modern Biology* 118
Genetics 118
First They Liked Planarians and Then They Didn't 121
Sages of Regeneration 125
Of Planarians and Genomes 132

8. *Planarians in the Popular Culture: The Arts, Science Fiction, Fantasy, and Humor* 137
Planarians in the Popular Culture 137
Planarian Man 139
Planarians and the Reimagined *Battlestar Galactica* TV Series 141
Fringe 145
Twilight 146
The Big Bang Theory 147
Dr. Who 147

PART IV | THE FIRST BRAIN

9. *The First Brain* 151
    Early, Really Early Nervous Systems 151
    The First Hunters 154
    The First Brain 159
    Why Are Planarians an Excellent Animal Model in Neuroscience? 164
    Very Brief Comments on Protopsychology 168

10. *From Corals and Plants to Planarians and Rats* 175
    Planarians in Pharmacology: Nicotine and Cocaine 175
    The Beginnings of Systematic Planarian Pharmacology Research 178
    The Temple University Team 180
    The Joy of Discovery 181
    From Corals and Plants to Planarians and Rats 183
    Planarian Research Translated to Vertebrates: What It Does Mean, and What It Doesn't 192
    What Can the Planarian Brain Teach Us About Our Own? 194

EPILOGUE 197
NOTES 199
BIBLIOGRAPHY 213
INDEX 231

Foreword

Robert B. Raffa, PhD

DR. PAGÁN IS a highly talented, dedicated, productive, and widely respected research scientist and teacher. I first learned of Dr. Pagán through his scientific publications. I have a long-standing interest in the use of planaria as a nonmammalian research model, and I was very impressed by his publications, in which he described several methodological innovations and novel applications using this model. His research and publications continue to impress me. I have also now come to know him and the absolute delight he takes in learning and writing about the world and our interactions with it. Dr. Pagán was trained at excellent academic institutions and programs. For example, his PhD dissertation work was done at Cornell, which is well known for its research in the field of neurobiology in general and drug abuse in particular, and was on the subject of chemical aspects of cocaine's action on dopamine reuptake transporters on brain neurons. He put this training to good use and has become a consistent publisher of his research and presenter at local, national, and international scientific meetings.

Dr. Pagán is a very active and engaged teacher in addition to his extensive activities as a research scientist. I know from my discussions with him and observations that he places teaching on a high pedestal. He goes above and beyond usual preparation and time allocation. He is even an active volunteer at "Science Day"-type events for young students. A passion to help and a caring for students at all levels are quite evident in any discussion with him.

Which brings me to this book. It is so much fun! It is written in a style that makes it equal parts educational, personal reflection, and motivational. The pure joy for the topic and the enthusiasm that exudes from the writing carry one along some well-traveled paths—but never, as one soon realizes, fully explored or fully appreciated until seen through his mind's eye—and some new paths. The surprisingly eclectic selection of subject matter, including from the arts, literature, and social sciences, can't fail to stimulate the reader's knowledge and logical (or sometimes illogical) thought processes. And don't be surprised if somewhere along the way one or more preconceived notions are challenged.

This is all presented in the most delightful manner. The style is not typical of most books written by such a distinguished researcher. It is, well, readable. And, dare I say it…fun. It is written in a way that is accessible to the beginner (yes, even to very young students), as well as more advanced students and anyone else who has even a modicum of interest in the world around us (and if you don't think you do now, you will after you read this book). Dr. Pagán masterfully uses the seemingly lowly planaria as a magical vehicle on this inspirational tour-de-force of life, liberty, and the pursuit of happiness in the appreciation of the natural world. Who would have thought that a book about the "first brain" could lead to such a wondrous stimulation and understanding of our own brain?

So remember back to when you were a child and you hadn't heard the word *science* yet. You were just curious about who you were, what was around you, what everything else was. So you looked, poked, or whatever else you did, to find out. Forget that this natural curiosity might have been lost, or suppressed, by school, by other demands, or by the seemingly imposing terminology of science. Want to get it back? Want to take up where you left off? Want to feel like a kid again? Then just read this book.

And, yes, it is okay to smile while reading a science book.

Robert B. Raffa, PhD
Professor of Pharmacology
Department of Pharmaceutical Sciences
Temple University School of Pharmacy
Philadelphia, PA, USA

Foreword

Harvey B. Sarnat, MD

WITH THEIR "CROSSED-EYED CUTENESS," planarians have fascinated me since we were first introduced to them in my high school biology class. As a university undergraduate student majoring in zoology, I continued my obsession with flatworms because of their extraordinary regenerative powers and the fact that they were the first animals to exhibit bilateral symmetry as a body plan, with the three body axes of all advanced invertebrates and vertebrates. My fascination with planarians became fixed when doing a master's of science degree in neuroanatomy and later, after medical school, when doing residency training programs in pediatric neurology and in neuropathology. I realized that planarians had evolved before the divergence of invertebrates and vertebrates and that, even if technically classified as invertebrates because they lack a notochord, many features of their nervous systems were more similar to those of vertebrates than of advanced invertebrates such as arthropods and molluscs. My ongoing research with the planarian nervous system continues to intrigue me as an origin of human life.

This popular science book of Dr. Pagán is important in bringing to public awareness the primordial role of the planarian in the evolution of animal life on Earth. From my medical perspective, their greatest practical importance lies in providing clues to regeneration that potentially might be extrapolated to the treatment of human organs and tissues that are defective in embryonic or fetal development, show progressive loss due to degenerative diseases, or become otherwise damaged by trauma, infection, exposure to neurotoxins, or lack of adequate oxygen or blood flow. No tissue of the human body has greater need of the lessons that planarians can offer than does the central nervous system.

The evolutionary origin of vertebrates remains a highly contested issue. Even if one accepts an origin from protochordates, it is unclear whether the lineage of cephalochordates (present-day amphioxus and its Cambrian period predecessor, *Pikaia*), urochordates or hemichordates, is ancestral. The relation of chordates to echinoderms is still debated because of some shared features with chordates. A free-living ancient planarianlike organism is a very attractive candidate as the ancestor of vertebrates including man, despite a few embryonic features that do not support this theory; for example, planarians are prostomes and vertebrates are deuterostomes.

No tissue of the body makes a stronger argument in support of the hypothesis that the planarian is a remote vertebrate ancestor than does the nervous system (see chapter 9). Planarians possess a bilobar brain with commissural connections and multipolar neurons with a single axone and dendritic spines, features of all vertebrate brains but rare among invertebrates. The paired neural cords are longitudinal axonal tracts surrounded by columns of gray matter neurons, a primitive architecture that persists in some parts of the human brain. Regrettably, molecular genetic studies to date have not yet been able to resolve the important evolutionary question of the ancestral origin of vertebrates in general and the vertebrate nervous system in particular. Because so many neurotransmitters are widely shared by many classes of invertebrates and vertebrates, their demonstration in primitive nervous systems tells us something of the origin of specific secretory types of neurones but not about evolution.

The planarian nervous system, if indeed ancestral to vertebrates, further offers an explanation to the mystery of why one side of the brain controls the opposite side of the body. To develop a functional *coiling reflex*, so that the animal defensively curls its body away from an aversive stimulus, it needed primary sensory neurones on one side of the body to perceive the potential threat and the discharge of motor neurons on the other side to contract longitudinal muscles on that side. This simplest of all reflexes required the development of the *decussating interneurone* to interconnect the sensory and motor neurone on opposite sides.

I consider it a great privilege and honor to have been invited by Dr. Pagán to write a foreword to his book and an opportunity to express an additional perspective as a physician dedicated to the nervous system and its disorders in the fetus and infant. I am proud to endorse this thoughtful and insightful book by Dr. Pagán, a unique contribution to bringing an awareness that the planarian is much more than a biological curiosity as a cute little animal with crossed eye-spots for the entertainment of high school biology students. The planarian is a missing link in understanding evolution and may offer a key to the treatment and reversal of many disabling diseases of humans.

Many individuals take pride in the fact that they can trace their family pedigree back several generations or even several centuries. I can boast that I trace my family origins all the way back to a pair of flatworms!

Harvey B. Sarnat, MS, MD, FRCPC
Professor of Paediatrics, Pathology (Neuropathology), and Clinical Neurosciences
University of Calgary Faculty of Medicine
Alberta Children's Hospital Research Institute
Calgary, Alberta, Canada

Acknowledgments

FIRST AND FOREMOST, I wish to thank my wife, Elizabeth, for her love, patience, and support, not only for this book, which she read in its entirety, giving me the most honest and useful feedback, but also because of our history together for the past twenty-odd years. Lisa, you are my dream girl, and I love you.

To my children, Vanessa, Reynaldo, and Andy, thank you for your love, for making me a proud dad, and for inspiring me to become a better man every day of my life. I love you.

Even though he is no longer with us, I want to thank my dad for so many things! I also want to thank my mom for everything. Your answer to the question I asked you when I was about four years old, "Did God create microscopes?" is still as valid today as it was forty-odd years ago. I also thank my brothers, Oscar, Alexis, and Luis, as well as my sister, Luz, for being part of a wonderful childhood. Alexis is also a gifted graphic artist and drew several of the figures of this book. He has been doing things like this for me since my undergraduate days; we make a good team.

So many people have generously and enthusiastically helped me in this adventure. I want to start by highlighting the contributions of two friends who encouraged me in this adventure every step of the way. These friends are Mr. Peter Cawdron, a fellow blogger, friend and established writer, and Dr. Robert Raffa, a fellow scientist, friend, and collaborator. They both read the entire book and gave me invaluable feedback. They made sure that whatever I wrote was understandable to the general public without compromising the science; furthermore, they oftentimes proposed useful stylistic suggestions that improved the manuscript. I also want to thank Dr. Raffa for suggesting the title of the book: *The First Brain*.

I decided to start blogging to "test the popular science waters" after reading a two-part blogpost from Mr. Brian Switek's *Laelaps* blog titled "So you want to write a pop-sci book?" Blogging made me think seriously about the possibility of writing a popular science book, but I actually decided to start writing it after reading an article by a professional scientist and science writer, Dr. Mark Changizi, titled "How to Write a Popular Science Book," which appeared in *Psychology Today*. From these readings I truly believed that I had what it takes to write popular science, so from the bottom of my heart, Mark and Brian, thank you.

I thank Dr. Sina Ebnesajjad, who gave me instrumental advice to navigate the book contract intricacies and for his enthusiastic support and friendship. Every time we met at the coffee shop he made sure that I was at the computer actually writing!

I thank Dr. Harvey Sarnat and Dr. Robert Raffa, whose scientific papers made me think about planarians in terms of neurobiology and pharmacology, respectively. Also, I thank them for the forewords to the book and for their unconditional support for this project.

I wouldn't have had the courage to go ahead and write a book proposal without the feedback and encouragement of many people that read my *Baldscientist* blog. Several fellow bloggers are regular readers who always gave me support and kind words, but most importantly, they gave me their honest feedback on my writings in general and on some of the book chapters in particular, with no sugarcoating whatsoever, for which I am immensely grateful. These regulars are Peter Cawdron, whom I mentioned above (Thinking SciFi), Sedeer El-Showk (Inspiring Science), John Jaksich (The Silent Astronomer), Sam Mason (Science in the Land), and Rebecca Trotter (The Upside Down World).

Next, I wish to thank Drs. Javier de Felipe, Harvey Sarnat, and Bruce Johnson for their feedback on the neuroscience chapters and Drs. Emili Saló, Francesc Cebrià, Mike Levin (who also provided me with very hard-to-find papers), Phil Newmark, and Alejandro Sánchez Alvarado for their feedback on the planarian chapters and/or for great conversations on the matter. Drs. Margaretha Gustafsson, Giorgio Venturini, and Antonio Carolei generously provided hard-to-find reference material and provided a historical perspective on planarian neurobiology and pharmacology. Many other researches contributed by furnishing reprints of their original scientific work, for which I am also grateful.

I thank my longtime friends, Eddie and Kaori Reyes, for translating some rare planarian scientific literature from the original Japanese into English.

I must make a special mention of two people that generously provided materials, insights, and professional perspective about the general field of flatworm biology. Dr. Masaharu Kawakatsu sent me quite a few rare scientific papers on planarians and some original photographs of *really* hard-to-find early work (pre-1700s) on planarians and graciously allowed me to use such pictures in the book. He also proofread the planarian chapters and gave me a unique historical perspective on "planariology," of which he is one of the renowned protagonists. His gentlemanly character, which came through even in email communications, and erudite and enthusiastic advice proved invaluable for the completion of this book. Dr. Kawakatsu also graciously provided many original color photographs of planarians, which are included in the color plates. Also, through him I found Dr. Vida Kenk, the daughter of one of the preeminent planarian biologists of the twentieth century, Dr. Roman Kenk, as well as an invertebrate biologist in her own right. She also generously provided materials and personal insights on her father's scientific life through documents and through a delightful telephone conversation.

My road to academia was not straightforward. I was blessed with many people that made it possible: Vesna Eterovic, Richard Hann and Pedro Ferchmin (Universidad Central del Caribe), José G. Ortiz (University of Puerto Rico) and George P. Hess, Susan Coombs and Robert Oswald (Cornell University); you know what you did for me, and I thank you.

I want to express my gratitude to my academic home, West Chester University (WCU), especially the Department of Biology and the College of Arts and Sciences. I am fortunate for working at an institution where I do not have to choose between teaching and research; I can excel in both. I am also thankful for WCU's fine library staff and resources. Libraries are still a fundamental scholarship resource, and I do not think anyone can convince me to feel otherwise.

At the publisher's front, my thanks go to Mr. Jeremy Lewis, Mr. Erik Hane, and Ms. Hallie Stebbins of Oxford University Press, who believed in my work and are my knowledgeable guides in the initial editorial process. My special thanks to Ms. Eli Hausknecht and Ms. Molly Morrison, project managers and to Ms. Danielle Michaely, copy editor. My special thanks to Laura Shelley, master indexer. Thank you all for making this "first book-writing adventure" a reality!

All these people helped me to create a much better book than the one I would have written without their input. Any mistakes, of course, remain my sole responsibility.

Finally, I want to thank my barber, Mr. Heath Brewer, who is always in the mood to talk science.

When we're in college, we think about our future as a direct line from now to then, from here to there. . . . But if you look at the careers of many successful people, you'll find that their route is often far more sinuous. And if you look at happy people, you'll find even fewer who traveled a straight line.

—LEONARD MLODINOW

INTRODUCTION

FIRST AND FOREMOST, thank you for reading this book! As an avid reader myself, I have always thought that reading is a little bit like a conversation—granted, a somewhat one-sided conversation, but in my mind, a conversation nonetheless. Moreover, what is lost in this type of conversation in terms of direct interaction is more than compensated for in terms of the number of people who can converse with the author, as well as by the elimination of the time factor. You see, when we read, we are in a sense hearing the voices of authors we may never meet in person for a variety of reasons, including that they may no longer be with us. With this book, for the first time I will be on the proverbial "other side" of the fence in this type of conversation. This is very exciting for me, so again, whether you are reading this right after publication or whether you are reading these words a hundred years from now and wonder about the childish state of science in my time, from the bottom of my heart, thank you.

I absolutely love science, always have, always will. I am in awe of the natural world and of the methodology of science that allows us to uncover nature's secrets. Nature is majestic; I know of no better way to describe it. I feel quite privileged and even grateful of being able to try to understand some of it, however little I can really understand in my limited time, ability, and opportunity on this planet. Like the great Isaac Newton, many times I feel like ". . . a boy playing on the sea-shore, and diverting myself now and then finding a smoother pebble or a prettier shell than ordinary, whilst the great ocean of truth lay all undiscovered before me."

On a related note, I feel incredibly lucky that I work as a scientist and educator. I hold an undergraduate degree in general natural sciences, a master's degree in biochemistry, and a doctorate in pharmacology, with a strong emphasis on neurobiology. There is a point in stating this. In a way reminiscent of the sentiment stated in Dr. Mlodinow's quote at the beginning of this introduction, I did not follow the direct route to pursue my academic dreams. You see, I was a nontraditional student. After college, I was a high school teacher and I worked in a casino (not at the same time) for a couple of years. After that, I got a job as a research technician at a medical school, where I worked for about nine years. I would say it was there that I learned how to be a scientist, for which I will always be grateful. I then got the opportunity to get my PhD at a top research university and now I am having the time of my life because I teach and do research.

I love my job. Do you want to know why? In what kind of job do they pay you explicitly for reading and learning about the topics you love? It gets better; as a science professor, a big part of the job is to talk (usually to students) about scientific subjects, and the best thing about that is that you have a captive audience: *they have to listen to you!* I am also a practicing scientist, which is a sometimes overlooked way of learning and teaching. After all, when you do research, it is like having a front seat while learning about nature directly from the source. As part of this, I run an active research laboratory in which I have had and still do have the fortune of interacting with many dedicated undergraduate and master's degree students who learn about science firsthand under my direction. I have been very fortunate in the sense that by far, most of these students have what, in my opinion, they are supposed to have to succeed in life: enthusiasm, responsibility, and a pristine work ethic (a good mind, while quite convenient, is just icing on the cake). That said, remarkably, without exception, each of my past and present students has taught me something. Along those lines, many of my past research students have gone on to develop very productive careers of their own, which understandably, makes me a very proud mentor.

Even though I am very familiar with the topics I will talk to you about and in many cases I even have direct, firsthand experience on the subject matter, keep in mind that in science, you are free to verify—actually, you are *encouraged* to verify—anything that any scientist says about science, no exceptions. Nowadays, this is easier than ever! Even more importantly, one of the most beautiful things about science is that all of us, whether we are aware of it or not, are innate scientists. Allow me to explain. Without exception, humans are curious creatures. In that sense, we are like many other organisms with which we share this planet. However, only we humans have built upon that curiosity by developing methods and ways of thinking that permit us to understand nature a little better, a little beyond what we can immediately perceive through our bare senses.

It is pretty evident that I am rather enthusiastic about this book; it is expected. However, at this point the most important question to you, my dear reader, is, *what's in it for you?*

Well, the first thing that comes to mind is that if you are interested in science, it is a safe bet that you are interested in its progress; you want to learn as much as possible of a particular subject, especially where it is going, while being careful at the same time of not losing sight of where it came from. You will have an easier time predicting a possible future trend or application of a specific discovery if you understand how it came to be.

Also, we science types are fascinated by facts, but facts by themselves are simply not enough; we also want to know whenever possible the "why" behind those facts. If you care about this—in other words, if you are even peripherally interested to find out about the "whys" of nature—you are implicitly interested in the organ that makes you feel interested. That organ is, of course, your brain.

In this book I will tell you a story with science, especially neurobiology and pharmacology, as the unifying theme. The book consists of ten chapters organized into four parts. What's different about this book is that its main character is a certain type of flatworm, commonly called planarians. Before meeting these interesting little guys, I'd like to offer you a brief introduction about science in general; after all, this is a popular science book. In part I, "Fundamentals," we will first go over a series of science-related concepts in two chapters that will make the understanding of the rest of the book much easier.

The main theme of this book is to explore aspects of nervous system anatomy, physiology, and pharmacology from the premise that the planarian brain is an early, if not the earliest, example of an actual brain in the animal kingdom. Moreover, we explore the concept of the planarian brain as the first vertebrate-style brain that appeared in our planet and how the study of the planarian brain may throw light on the human brain, helping us understand ourselves a little better.

Before we do that, though, we must have a working understanding of neurobiology. Part II of the book, "The Science of the Brain" (chapters 3, 4, and 5), explores the neurobiology- and pharmacology-related concepts that are necessary to understand the rest of the book. These topics include the similarities and differences between "neurobiology" and "neuroscience," among other topics. In this part we will also talk about what is often called the most complex system in the known universe, the human brain.

In part III, "Planarians," we finally meet the main character of this book. In chapters 6 and 7, I will formally introduce you to some of my favorite (nonhuman) organisms: the *flatworms*, specifically the *planarians*. These worms have a distinguished place in developmental and regeneration biology; you may remember them as the little guys that grow a new head when decapitated (more on this later!).

Planarians have captured the imagination of people of all walks of life, not only scientists. In chapter 8 I offer you an interlude in which I explore a series of curious connections between planarians and popular culture. Just to give you a preview, these guys have inspired works of art, they have appeared as characters in comic books and other kinds of literature, and there are even some interesting links between flatworms and very popular science fiction and fantasy works.

Part IV, "The First Brain," is kind of the "flagship" of the book. It is also about neuroscience, but exclusively from the perspective of planarians. In chapter 9, we will see that bits and pieces of what constitute the modern nervous system are found in much simpler living organisms. Also, we will explore in more depth the case for the planarian brain as the first actual example of a brain, as well as some intriguing neurobiology research done in these flatworms.

Psychology can be seen as applied neuroscience. One of the most interesting episodes of the history of psychology also has planarians as the main characters. In this chapter I also explore this interesting history, where planarians were for a while a favorite animal model to study learning and memory. In fact, for a very brief period a new discipline, "protopsychology," did show a lot of promise, was intensively studied, and then simply faded away, in that order.

More recently, certain species of planarians have become a favored, popular, and somewhat surprisingly useful animal model in pharmacology and toxicology. We deal with this emerging "planarian pharmacology" field in chapter 10. Among other topics, I will talk about the experiences of several fellow planarian researchers, some of whom are also colleagues and friends. More specifically, we will talk about the effects of naturally occurring products on drug-induced behavior using the planarian model. Interestingly, several research groups, including my own, have uncovered a series of parallels between planarian pharmacology and vertebrate pharmacology, which provides evidence in favor of the usefulness of these invertebrates in pharmacological research.

Whenever I talk (or write) about science, I try to stress the fact that science is done by people, with everything that goes with the territory. In any scientific discovery the human factor and its historical background are factors that need to be taken into account to understand the full story. For this reason, in all chapters I have tried to include historical and character anecdotes that help present the human factor in science.

In this vein, I want to point out that each chapter has a number of (not too many, I hope) endnotes in a separate section at the end of the book. I encourage you to take a look at them. They will not only oftentimes point you in the right direction for further reading but also clarify and put in perspective the discussed topics. In some of them I even included humorous anecdotes and miscellaneous tidbits related to the topic at hand. Also, by reading the notes as you read the text, it can help you go around a problem that I myself have. You see, all of us were raised in the era of television. This means that most of us are accustomed to paying close attention for a few minutes and then being interrupted by commercials. Many of the notes will serve this purpose by shifting your attention ever so briefly so you can go back to the text with refreshed concentration.

The end of the book features an epilogue that tells the fictional story of two characters (one historical, one imaginary) linked by a single planarian over more than a hundred years.

I would say that there are two main purposes of this book. The first one is to share with you my fascination with this subject matter. The second is to inspire you to learn more about the topics you will read about, to go beyond what I say here and hopefully enrich your mind in the process. In this respect, I will try to follow the distinguished example of some of my favorite writers/scientists, including Stephen Jay Gould, Carl Sagan, Richard Dawkins (when he writes about science) and, of course, the author who introduced me to the popular science literature in the first place, the one and only Isaac Asimov. I hope you find my writings at least a fraction as entertaining and inspiring as these authors' works have been to me.

I wrote this book with interested laypeople in mind; however, I would be very happy if biologists and other scientists find something of value in it too. My hope is to try to make my readers curious enough that they may want to expand their knowledge in these and other related areas. To help you go beyond the scope of the book, I have included notes and selected references for each chapter as I said earlier, as well as a general bibliography. Each chapter can be read independently, but I suggest you read them in order, as I often refer to previous chapters, depending on the topic. In general, when I introduce a concept for the first time, I will define and/or explain it. Some chapters include material and original ideas that first appeared in my blog, *Baldscientist* (http://baldscientist.wordpress.com). However, any material from the blog is significantly expanded and updated.

I want you to be aware of something before starting this journey with me. No single book, especially a popular science book, can say all there is to say about any particular subject.

I have tried to do justice to the wealth of information about planarians available in the scientific literature. This has been a process full of difficult choices, as there is simply not enough space in a single "normal size" popular science book to include a comprehensive account of these interesting organisms. In all fairness, this is true of any life form in this wonderful planet of ours. That said, in no way am I implying that the contributions that did not make it into the book are of lesser importance than the ones that did. Planarians are experiencing a true renaissance as important animal models in biology. There are many laboratories led by brilliant scientists working on various disciplines, including but not limited to molecular and developmental biology, biochemistry, regenerative biology, neuroscience, and pharmacology. These research groups are continually generating data and adding important findings to the field. I want to apologize in advance to all my scientific colleagues whom I do not explicitly mention in this book due to space considerations, both those I know personally and those I know by reputation.

Now, would you like to know about the first planarian I ever met?

Life can be full of surprises and amusing coincidences. As I was in the process of researching and writing this book, I remembered an episode of my childhood that describes the very first time I ever saw a planarian.

It was around 1975; I was ten years old and, like many kids, loved to play outside, an activity that sadly, for many complex reasons, is not as common anymore. My elementary school yard was huge, or at least I remember it that way. When it rained hard, a couple of small ponds usually formed, which were rapidly colonized by the wonderful critters that children love to play with. Alongside my friends, I loved to collect tadpoles; they were "my thing." Sometimes I even brought them home and watched them for several days as they sprouted legs and eventually became small toads who happily hopped away.

I vividly remember this one time when I was looking for tadpoles and I saw this whitish, ribbonlike wormy. I took it to my classroom and showed it to my science teacher. We put it in a small fish tank we had in our classroom. In retrospect, that was not a good idea. I could not find it in the tank afterwards; I am sure that a fish ate it. I do not remember if it was my teacher who told me that the wormy I brought her was a planarian or whether she gave me a book so I could try to find out what it was by myself, but at any rate, it was right then and there that I learned about planarians for the first time.

HENRY THE PLANARIAN.
Illustration by Nathan Scheck (http://www.albinokraken.com). Used with permission.

I never imagined that thirty-odd years after my first encounter with a planarian I would be doing scientific research with these wonderful beings, much less writing a book about them! I hope you enjoy reading this book as much as I enjoyed writing it.

I

Fundamentals

All our science, measured against reality, is primitive and childlike, and yet it is the most precious thing we have.
—ALBERT EINSTEIN

1

SCIENCE

SCIENCE DOES NOT EXIST

SCIENCE DOES NOT exist. This is a strange way to start a science-themed book, isn't it? There are several reasons to say this. The main one is that science is a human process that lets us divide nature into manageable chunks so we can understand it better. Nature is a continuum, which means that there are no definite frontiers. Sometimes, especially in modern biology, it is very difficult to decide whether you are studying an organism in terms of biochemistry, physiology, or pharmacology, among many other disciplines, as all these areas of knowledge are seamlessly integrated. More on this soon, but I have to say that in our times, it is practically impossible for a single person to learn in detail all aspects of our current understanding of nature. In the past, this was easier, simply because there was way less information about the world, which brings us to the second reason I say that science does not exist.

Up until about two hundred years ago or so, the study of nature in general was collectively called *natural history* or *natural philosophy*. Similarly, the people who occupied themselves with natural history were called *naturalists* or *natural philosophers*. Today, we call these people *scientists*. In those times, natural philosophers generally studied all of nature, at least as they understood it. Sometimes they even studied it without a clear distinction between the animal, vegetal, or mineral world and it was indeed possible for a single person to have a working understanding of pretty much all the science of the day. Moreover, the word *scientist* did not even exist up until 1835, when it was coined by the philosopher William Whewell, who suggested it as an analogy with the word *artist*.[1] As you may imagine, in those times you did not call yourself a scientist and could not even decide to "become" a scientist, as science was not an actual formal occupation. Natural history was usually an activity done by people who had the free time and financial resources to dedicate themselves to study nature. They were only able to do so because of their wealth, with the financial backing of royalty or other people of independent means, and even by joining the clergy. Also, it was not uncommon for a naturalist to raise money by selling plant, animal or mineral specimens collected in their expeditions to collectors, for example. Essentially, science was a hobby, but what a hobby! Regardless of the means of support, to do science in those times was a true labor of love. There were other occupations where you could earn a living with much less effort. In those times, whoever did science did it because he (and sadly, in those times it was usually a "he" as opposed to a "she") wanted to.

The development of the natural sciences gave humanity a whole new series of points of view. As we said, science as an activity is a relatively recent human construct, a method and a way of reasoning, by which we try to make sense of nature. In these modern times, it is critical to understand science for two reasons. First, we live in a highly technological society; as time goes by we are becoming more dependent on the practical applications of science and yet, many people have very little idea (if any idea at all) of what science is and how it works. This is a rather dangerous combination, because ignorance about science can lead to uninformed decisions, especially by people in authority, which can have disastrous consequences.

Second, now more than ever, the boundaries between the traditional sciences are not very well defined. For example, two hundred years ago it was perfectly possible to study living organisms (biology) without knowing any significant amount of chemistry, physics, or mathematics. The main reason for this is that the life sciences were mainly descriptive as opposed to experiment oriented. In those times, especially in the life sciences (barring some notable exceptions), naturalists were primarily concerned with the "what it is," sometimes with the "why it is that way," and rarely, if ever, with the "how it got that way." In our days, we would say that natural philosophers were concerned with the identity of something and its purpose in nature, real or perceived, and for the most part, with some distinguished exceptions for sure, naturalists were not really concerned about what made things tick.

This approach to the natural world is perfectly fine if you just want to admire nature and its sheer beauty in a contemplative manner. Many people do this and that's perfectly okay. In fact, most scientists start in this way. Pure and simple curiosity ignites the proverbial spark that triggers the interest of a person about anything. This, in turn, brings about the sense of wonder—nothing more, nothing less; this is part of human nature and scientists are no exception.

On the other hand, to do significant, reliable, and applied science (in other words, to actually put the contributions of human knowledge to good use), this contemplative frame of mind in itself is not good enough anymore. We now have access to a whole series of deeper levels of understanding of nature that are only possible because of the application of theoretical and methodological tools that enhance the limitations of our senses. The development of these tools was only possible through a better understanding of mathematics and of the physical and chemical sciences.

Happily, this deeper understanding of nature is just as beautiful as, and maybe even a little more beautiful than, what is immediately apparent with our unaided senses. You see, we now know that life is a process intimately related, even identical, to chemical events; this fact in itself is truly remarkable. In turn, chemistry is essentially physics, and we can only understand physics through mathematics; there is simply no other way. This is why mathematics was famously declared to be the "language of nature" by none other than Galileo Galilei, the original scientist's scientist. It is only fair to point out that many consider mathematics, as well as science, as just examples of human constructs; this is a hotly debated point. Was mathematics discovered or was it invented? If it is the former, it would mean that we can think of mathematics as an integral part of the universe; if it is the latter, it would mean that it is a product of the human mind.

Now I need to say something that I will say several times in this book in one way or another: very few things in nature are absolute. This is part of the beauty of the universe. In those lines, I think that mathematics was at the same time both discovered and invented; it only depends on how we look at it. Regardless of anyone's opinion, the truth is that mathematical ideas that were thought to be of purely theoretical or abstract value at the time when they were invented (or discovered) were later found to have an actual physical reality in our universe.[2] This has been demonstrated time and again. Some have even said that the "…enormous usefulness of mathematics in the natural sciences is something bordering on the mysterious…."[3] It is undeniable that mathematics is a very useful way (sometimes the only real way) to describe many aspects of the natural world. You may be wondering why I am talking so much about mathematics in a book ostensibly about biology. Well, mathematics, as applied to the life sciences, has rightfully been called "biology's new microscope, only better."[4] This is rather interesting, and if you think about it, you will realize how much sense it makes. It is very clear that the invention of the microscope allowed us to examine life in an unprecedented level of detail; we were able to see aspects of nature that we had no idea existed, including life forms that had never been observed before. In a similar way, mathematics shows us aspects of the universe in general and of the living world in particular that were unimagined until now, opening a true new world of possibilities. Therefore, science and her big sister, mathematics, as well as their little sister, technology, are human activities that help us explore nature. After all, ours is the only kind of technological mind that we know of, so far.

It is very important to point out that when I say the processes of science or mathematics are human constructs, in no way am I implying that whatever they help us discover about nature is a matter of cultural or personal opinion. As we will see shortly, our interpretations about anything that may be observed in nature can and do change over time; this is a normal—essential even—part of the scientific process. This fact, however, does not negate the *reality* of a natural process. Nature chooses to behave in a certain way, independently of how we decide to explain it, period.

One of the most awe-inspiring aspects of nature is that it is a continuum that goes from the subatomic realm to the phenomenon of mind itself. Not surprisingly, these two extremes are arguably the least understood aspects of nature; to some, they are the most interesting, but of course, we can only see things from our own perspective. As we saw before, one of the most useful aspects of science is that it allows us to virtually divide the universe into manageable blocks that lend themselves to closer examination. The main idea is that once we observe and describe a certain number of such blocks, we combine the data derived from many of such observations into a logical model that can help us to understand the whole a little better. We call this approach *reductionism*.

REDUCTIONISM AND ITS ADVANTAGES AND LIMITS

In the reductionist approach, one starts by looking at small fragments of the whole. This concept has been around since the times of classical Greek philosophy, but the general consensus is that reductionism was explicitly defined for the first time in the seventeenth century by the famous philosopher René Descartes.

How do we apply the reductionist approach? More importantly, why would we want to? In a nutshell, reductionism allows us to divide any system of the world into manageable chunks. For example, how would you describe the parts in Figure 1.1?

To describe part A you have several options. You may choose to list all the sizes, shades, and markings in the picture, or say something like "many round shapes of different kinds," for example. In any case, to describe part A you will

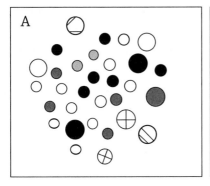 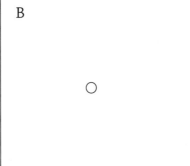

FIGURE 1.1. A. Complex system. B. Simpler system.
Illustrations by O. R. Pagán.

need to invest more time and effort than you will to describe part B. In this case, you will simply need to say "a circle" or something along those lines. The idea is that if we are able to describe in some detail a small part, say, one small circle, we hope to be able to extrapolate and apply whatever information we get to a much bigger system, like in the more complex picture. In other words, a simpler system is easier to study than a more complex one. Reductionism is far from perfect and sometimes is very difficult to apply. There is no guarantee that we will gain a complete understanding of a complex system just by examining its simpler components. However, we have to start somewhere, and frankly, reductionism is the best we can reliably do right now; despite its limitations, it has proven very useful to understand our universe.

That said, it is important to point out that the reductionist approach, as powerful as it is (warts and all), does not, will not, and cannot tell the whole story about a particular aspect of nature. It is a little bit like trying to understand, say, a bicycle just by studying its component parts separately. The study of the separate parts is necessary, but not sufficient, to really understand how it works. It is only when you figure out how all the parts work together that you will really understand the whole. We can easily apply this example to more complex types of machinery like cars, computers, or even brains. I will elaborate on this concept soon.

There is an additional feature that is often present in many complex systems, the phenomenon of *emergence*. Simply stated, an *emergent property* is an occurrence that is not immediately apparent, as it is not possible to explain or even predict by just examining a small part of the whole. In other words, this is essentially the old concept stating that the whole is more than the sum of its parts, just taken to the next level.

Let's illustrate this with a somewhat silly example and we will then work our way up. For example, how do things get wet? We all know that if you pour water on something, it will probably get wet, right? Now, my question is, can a water molecule get wet?

No, it cannot. You see, the general phenomenon of "wetness" only appears when a huge number of water molecules coat the surface of an object. It is a combination of a series of physicochemical events in which a film forms when water molecules interact with each other through a property called *cohesion* while simultaneously interacting with a surface through a property called *adhesion*. It is simply not physically possible to get a water molecule wet.[5]

Now, the *perception* of wetness is an altogether different thing, even though perception itself is also considered an emergent property! In this case, the events starting when your arm gets wet end up as the feeling of wetness, a perception that happens only in your mind.

Another way of looking at emergent properties is by thinking of that frequent companion of intellectual thoughts, a cup of coffee. Very few of us think about the coffee that we intend to enjoy after we prepare it and season it the way we like it.

Can you keep a secret? Are you sure? Here it goes: the flavor of your coffee has no physical reality at all, none whatsoever. Allow me to explain.

When you are about to take that first sip, many factors come into play that contribute to your perception of how it will taste. The first thing you perceive is the aroma, which is integral to the flavor experience. Other factors include temperature and the many chemicals that are released in the brewing process. It's not just caffeine. There are at least two hundred compounds that contribute to coffee's flavor and aroma yet some estimates say there are up to eight hundred different classes of compounds in an average cup of coffee! Also, any seasonings you may have added to it like sugar, cinnamon, honey, and so forth do contribute to the overall experience. Your nervous system detects, combines, and interprets all these chemical signals; this interpretation leads to the creation of the subjective flavor perception, which, again, happens only in your mind. Flavor is subjective precisely because it is a literal matter of taste; for example, the exact same cup of coffee can taste delicious or dreadful depending on who's drinking it.

There are many other examples of emergent properties in nature. Consciousness, as we saw previously in the example of touch perception (wetness) and taste perception (coffee), is closely associated to our very essence as thinking entities. At the moment, despite what you may have heard, nobody seems to really know *exactly* what consciousness is, let alone how it is generated; there are a lot of opinions, discussion, and confusion about this, but the general belief is that consciousness somehow arises from the multiple, complex interactions of billions of nerve cells in the brain. We will talk more about the brain's complexity in part II of the book.

Finally, life itself seems to be an emergent property! It seems that life appears when a critical number of molecules of a certain type organize themselves in a particular way. In other words, a molecule cannot be alive; a whole bunch of molecules just mixed with each other in an unorganized way will not be a living thing either. Rather, a big—really big—number of molecules, organized in a particular way, as in a cell, are alive (as long as the cell interacts with the external environment). We will see aspects of this in the next chapter.

Before that, let's look at this from a different perspective. A carbon atom within the lump of coal you may have received from Santa for being naughty is indistinguishable from a carbon atom within the cute puppy your neighbor received at Christmas because she was nice. What makes a puppy and not a lump of coal a living being are not the atoms or molecules themselves; instead, what's important is how these atoms and molecules are organized with respect to other atoms and molecules and how that organization allows information to be exchanged with the external world. By the way, the concept of organization originates the word *organism*.

The perils of overemphasizing the reductionist approach is illustrated by the trend, starting in the 1950s to 1960s, of taking the philosophical stance stating that true biology could only really be studied from a molecular perspective. This misguided (but understandable up to a certain point) way of thinking was highly influenced by the rapid development of powerful molecular biology techniques, which were improved over time in an almost exponential way, a process still happening to this day. Another aspect of the misguided dismissal of the "old biology" was the personality of many of the protagonists in this chapter on the history of science. In many of these characters we can really see the good, the bad, the ugly, and the just plain rude. Many of the molecular biology groundbreakers wrongly dismissed other aspects of biology, like taxonomy, for example, calling it mere "stamp collecting," to borrow an expression of one of physics's great minds, Ernest Rutherford, who considered physics the only "real" science. James D. Watson, one of the discoverers of the structure of DNA, emphatically championed this dismissive attitude toward "classical" biology. Amusingly, "classical" biologists like Edward O. Wilson and Ernst Mayr magisterially refuted these "new biology" arguments with erudition, class, and above all reason, but this is a story for another time.

At any rate, this way of thinking created an artificial and arbitrary distinction between molecular and organismal biologists, a distinction that is happily disappearing from modern biology. For example, organismal and evolutionary biologists as well as ecologists are increasingly using molecular biology techniques, and conversely, many molecular biologists are thinking in terms of the evolutionary and ecological implications of molecular biology discoveries.

Many biologists (including yours truly) are very happy with this gradual reintegration of these biological subdisciplines, as the true significance of a molecular event can only be properly understood in the context of the whole organism and its interaction with the environment. I am always very much amused when I hear some scientists proudly talk about the identification of a new receptor or the sequencing of a gene and even about sophisticated ways to measure gene expression *as if those were the main objectives in themselves.*

This happens more than people realize; many scientists truly fail to see the forest because they are looking at the trees. Don't get me wrong; discovering a new protein or sequencing a gene is good and essential to scientific progress in biology. We really do have to understand the parts if we want to have any hope of understanding the whole; there is no other way. The thing is that, in itself, the discovery of a gene or a protein or even determining the association and dissociation rate constants of a drug acting on a specific receptor (I couldn't help myself; this is part of my own field and it is wonderfully interesting—look it up) is meaningless outside of its physiological context.

Again, in essence, if true progress is to be made in scientific matters, you must eventually have to take a look at the proverbial whole picture. There is an increasing body of work about "complexity science," which tries to explain nature and emergent properties by integrating the multiple parts obtained from a reductionist point of view into a coherent model; there are several books that clearly yet masterfully explore this topic.[6,7]

Let's now take a look at a couple of practical examples of reductionism.

Suppose you want to know as much as you can about, say, a car. The first obvious thing you can look at is its physical appearance. For example, you can describe its color and general shape. Some of these aspects, like color, can be shared by many types of cars, but the shape of the car is usually make specific. You can even obtain blueprints to look at how the internal structure of the car is organized. Again, some of these aspects are shared with other cars, while others may be particular to each particular make. Next, if you are mechanically inclined, you can try to figure out what makes the car tick, how all its parts work together in a precise way so that the car will move, how the signal lights work, what kind of fuel it will need, and so forth. Also, if you like engineering, you can make a list of all the parts, down to the smallest nuts and bolts, and how they all relate to each other. Almost nobody thinks about the specific materials that make up the parts of a car, but if you are curious about it, you can find out which parts are made out of aluminum, iron, copper, and even plastic or leather.

As you can see, there are many layers of complexity in any multipart system. Now let's look at a biological example.

The building blocks of a typical brain, as in any biological organ, are its cells, including its neurons and glial cells, which we will talk about soon. Suppose that you wish to study a nerve cell and want to learn as much as you can about it. As with the example of the car, we can first try to describe this cell through several points of view; again, many parts will be found in most cells, while some others are specific to neurons and sometimes even only to certain types of neurons. This cell can be observed through a microscope at different magnifications using one or more of the currently available microscopy techniques. Using these techniques, we can begin to understand the neuron from the perspective of what it looks, for example, by looking at its surface and by observing its physical shape (in other words, its *morphology*).

We can go further, by working our way down to smaller parts, also known as its *ultrastructure*. We can even use special fluorescent markers to label specific cell components; this technique makes gorgeous pictures. When we describe a neuron using the techniques outlined earlier, we are looking at the cell in terms of its *structural biology*. In an analogous way, if we were describing the parts of an organism, we would talk about *anatomy, histology,* and so forth.

Once we have information about this neuron's structural features, we could also examine the various processes that make it work, as described by its *physiology* (anatomy is concerned with structure, while physiology is concerned with function and mechanism). We can start from the multiple transport phenomena that provide the cell with nutrients to the ion channels that open and close in a highly coordinated manner (in the absence of any pathologies) to allow the neuron to communicate with other cells. In turn, groups of neurons coordinate many processes that eventually generate behavior, including consciousness itself.

Next, we can describe this neuron in terms of *molecular biology and genetics* by looking at its genome (the genetic complement of an organism; in other words, the ensemble of all its genes) and by figuring out which genes are on or off at any specific point in the course of its development. One of the most intriguing questions in all of biology is: How does the structure and function of a living organism arise from a series of gene-coded instructions? In other words, how can a living being, say, a person, start as a single cell and grow into a body containing *trillions* of cells of many different interacting types? How does that happen? We have a lot of information about it, but still, nobody knows precisely the mechanics of it. However, we know one thing for sure, the process of *development* is essentially *genes in action*.

Going back to the neuron, we can go beyond molecular biology and figure out the neuron's basic chemical components and how they are organized into macromolecules (macromolecules are big molecules; more on that soon) that, by virtue of their interactions with many other molecules, are able to capture energy from the environment and use it to enhance their chances of survival, in other words, to display the phenomenon of life. This would be classified as *biochemistry and biophysics*. In principle, we could go on and examine this cell at the subatomic level, but life is poorly understood at this level anyway—I think you get the idea.

Essentially, in the example of the car, even if we get a thorough list of all its parts, that does not tell us much about how a car works. Similarly, in the example of the brain, again, even if we get a complete listing of all its component parts, that does not tell us much about how the brain works. It is only when we gain at least a working understanding of how all those multiple parts work together and how they relate to their environment that we achieve a more complete intellectual grasp of the complete entity.

We could go on listing all the possible ways we can use to try to understand the neuron. However, do you think the neuron cares?

A neuron exists and does "its thing." That's it. A single neuron is not aware or conscious in any sense that we can recognize as such. In fact, it has been said that "None of your neurons know who you are, and they don't care."[8]

The neuron exists; nature exists. Science is simply an intellectual tool, albeit an excellent and powerful tool, that we use to understand nature.

Throughout this book, we'll talk about different sciences, mainly neurobiology and pharmacology. These are scientific disciplines that began to be developed as independent subjects within the general sciences over the last two hundred years or so, as knowledge advanced through observation and experimentation. Once again, it is important to keep in mind that in reality, science includes all the different ways to describe life, the ultimate emergent property. Let's explore some more important concepts about this.

ON THEORIES AND THEORIES

In any conversation about biology, the concept of evolution is there, explicitly or implicitly. The idea of evolution is arguably the most fundamental concept that helps us understand the living world. However, despite its evident importance, the concept of biological evolution is the most abused, misunderstood, and maligned term in all of biology, bar none.

One of the most commons ways in which evolution is dismissed is by calling it a mere *theory*. Well, in all fairness, evolution is indeed a theory; or rather I should say that evolution is a *Theory*.

Confused?

You are not alone. This confusion is brought about by the usual lack of a clear distinction between the two meanings of the word *theory*. A *theory* (uncapitalized) is essentially a guess, an opinion, as in "This is how I think it happened...." In this sense, a theory does not have to prove or explain anything. It is merely an opinion, nothing more, nothing less.

A *Theory* (capitalized) in the scientific sense is a logical model that collects and organizes a wide variety of phenomena, integrating them in a coherent framework. Also, a Theory does not depend on personal opinions. When thought about in this way, a Theory can be useful to relate and connect a wide variety of observed phenomena and has the power to explain and relate additional observations that may be obtained at a later point. Most importantly, a Theory has the capacity to make *predictions* about things that may not have been observed yet. In turn, these additional observations can be used to modify the Theory or, in some cases, even change it altogether. This is the normal, tried-and-true, universally accepted method in the natural sciences. Make no mistake; it is accepted *because it works*.

A classic example of the predictive power of scientific Theories comes, not surprisingly, from Charles Darwin. In an 1882 book,[9] he discussed some of his observations of a certain species of orchids native to Madagascar, named alternatively the Comet Orchid or the Darwin Orchid (*Angraecum sesquipedale*; we'll talk more about scientific names in chapter 2). The Darwin Orchid belongs to a series of plants that are pollinated by moths. In some types of flowers, including *A. sesquipedale*, one of the petals is modified into a long spur, which is a hollow tube that contains a small amount of nectar at the very end. It just so happens that in the Comet Orchid, the spur is between eleven and twelve inches long! Furthermore, in these flowers, the nectar is only found in the last two inches or so of the spur. How can the nectar be reached by a moth? Based on this, Darwin hypothesized: "...in Madagascar there must be moths with probosces [sic] capable of extension to a length of between ten and eleven inches!"[10] It is important to realize that at that point in time, such a moth had not been discovered by science. Darwin then proceeds with his prediction, "If such great moths were to become extinct in Madagascar, assuredly the *Angraecum* would become extinct."[11] Since the orchids were still around, he was actually predicting that such moths did still exist. It was not until about twenty years after Darwin died that his prediction was proven correct. In 1903, a type of moth with a proboscis close to eleven inches long was found in Madagascar. This species of moth was subsequently observed to feed on the Comet Orchid and therefore was named *Xanthopan morgani praedicta*. The subspecies designation *praedicta* ("predicted") is a direct nod to Darwin's insight. This is one of the earliest and most interesting biological examples of the predictive power of scientific Theories. Who says that evolutionary Theory cannot make predictions?

Anyway, back to theories and Theories.

Things fall toward the ground. This is an undisputed fact that can be verified by everyone and happens each and every day. We usually describe this as *gravity*, and we can come up with many different explanations of why it happens. In fact, since antiquity, there have been several explanations (Theories) to account for the fact of gravity, from the time of Aristotle all the way up to Newton and Einstein and beyond in more recent times. And yet, despite all the possible explanations and models on how gravity works, the proverbial apple does not care about the mechanism; it falls toward the ground regardless of who is looking at it, Macedonian, British, or German. Therefore, even though we have had various Theories of gravitation at different points in history (and we may even have others in the future), we do not think of gravity as something dependent on our personal opinions or general consensus. Gravity happens; no amount of philosophizing can change this reality. In this sense, gravity is both a fact and a Theory, depending on what we are talking about. Now, speaking about evolution...

EVOLUTION AND LIFE

We can apply to evolution the same reasoning we used about gravity. The incontrovertible fact is that life in our planet has changed over time. This change has been thoroughly documented by many lines of evidence. As we go back in time in the fossil record, many different kinds of life forms are discovered. In most cases, these life forms are extinct. In this sense, evolution is a fact.

On the other hand, as with gravity, throughout human history several Theories of evolution have been proposed to explain this change over time through a variety of different mechanisms. Currently, the best available evidence points to *natural selection*, originally formulated by Charles Darwin, as the most likely mechanism that accounts for evolutionary change. There are other mechanisms that are thought to contribute to evolution, like *genetic drift* and *symbiosis* (the mutually beneficial close interaction of organisms belonging to different species). Even the fusion of whole genomes has been proposed as a mechanism for evolutionary innovation. The nature of science, particularly biology, is rarely black or white; there are many shades of gray. For example, we just mentioned that the very concept of biological evolution through natural selection is usually associated with Charles Darwin. However, it is only fair to point out that many other thinkers during history proposed comparable mechanisms with various degrees of similarity, correctness, and evidence to the one Darwin masterfully articulated. That said, Darwin's vision was the most coherent and complete, as evidenced by its lasting standing in science, a status conserved even today. Probably the one piece of information missing from Darwin's theory was the actual mechanism for biological change. With time, as we learned about genetics, we incorporated that knowledge into evolutionary thought.

An important point is that any given mechanistic explanation for evolution does not have to exclude another; sometimes, multiple mechanisms are equally right depending on the specific conditions, and even in some cases two or more distinct mechanisms can be complementary, again, depending on the context. I, for one, think the right mechanism to account for evolutionary change might just be a combination of several factors. My interpretation of this and other facts, of course, may change at any moment with the discovery of new phenomena; this is one of the many reasons science is so exciting—you never know what the future may bring!

Nonetheless, my opinion about the actual mechanism of evolutionary change is merely that, just an opinion, nothing more, nothing less. You see, I am a pharmacologist/neurobiologist with just a tad of physics envy, not a trained evolutionary biologist. This means that from the perspective of evolutionary biology I am an interested layperson—granted, a layperson fascinated with the evident evolutionary complexity of the living world, but a layperson nonetheless. This fascination makes me in no way an expert on the matter and, in that sense, again, my opinion about evolution is just that, an opinion. Let me give you another example to further illustrate this point.

In many circles, some of the arguments against evolution involve the opinion of "experts." The thing is, in more than one case, those experts are experts in areas other than evolutionary biology. It is like asking a podiatrist about how to treat a toothache. There is nothing wrong with podiatrists; it's just that every time I have a dental problem, I will talk to a dentist, and if my foot hurts, then I will see a podiatrist, thank you very much.

Furthermore, even if I were a trained evolutionary biologist, my personal opinion would not matter. You know why? My opinion—in fact, anyone's opinion—about the specific mechanism for evolutionary change would be beside the point, as it does not matter one bit in terms of *whether evolution happened or not*. Any opinions about what generates evolutionary change, however scientific (and trust me, properly trained evolutionary scientists have argued and will keep arguing about the specific mechanisms of evolution with various degrees of civility), do not change the fact that evolution occurred and is still occurring. Evolution is happening right now as you read this book.

In other words, these opinions or points of view have no bearing on the *fact* of evolution. Life arose on our planet; the multiple forms that life has taken over time have changed. That is evolution. End of story. A particularly lucid explanation of how the term *evolution* can be taken to mean both a Theory and a fact depending on the context used was given to us by the late S. J. Gould.[12] I encourage you to read it; you will not regret it.

Now, back to evolution. Briefly, evolutionary changes via natural selection from the perspective of genetics can be visualized more or less as follows:

- All organisms produce more offspring than can possibly survive (this means that since resources are not unlimited, under these conditions not all individual organisms in a population will survive).
- The population of a given organism is genetically varied (this means that there will be genetic differences between organisms of the same species). These differences will not only result in different morphologies (appearance) but also likely result in differences in biochemistry and physiology, for example.
- The genes present in an organism (not the species) will influence survival. For example, organisms may display pigmentation that allows them to blend in with their surroundings. This may allow them to hide from predators or prey. These organisms will tend to survive for longer, increasing their chances of reproduction. This means that surviving organisms will have the opportunity to leave descendants, which is, after all, the main point of biology.
- Eventually, the genes that allowed the organisms to survive will become more prevalent in the population, since only the surviving organisms will be capable of passing on their genes.
- This implies that, over time, the population as a whole will display a higher proportion of their own genes, not only the genes that provided (by chance) the organism a better ability to survive, but also many other

genes that were passed along as well (this is what we mean as change over time, a.k.a. evolution in action).

This scheme is sometimes called *survival of the fittest*, which is yet another misunderstood term related to evolution. This phrase has been called a tautology (basically, a circular reasoning) and used as an argument against evolutionary ideas. The tautology argument is a deeply flawed one for several reasons. The "survival of the fittest" phrase is rarely, if ever, used by professional scientists. Nevertheless, the fittest are indeed the ones that survive. What is going on? You see, the main confusion about this concept is that an important detail is often unmentioned; fitness is determined by survival based on genetics *within the context of the environment*. It is entirely possible that an organism "fits" in certain circumstances but not in others. Also, it is extremely important to point out that the survival of the fittest does not necessarily imply that the "fittest" are the strongest, the fastest, and so forth. Sometimes this is true depending on the specific case; however, let's think about ourselves for a moment. Humans are by no means the biggest, strongest, or fastest form of life—not even close. We do not have fangs, clawed paws, strong jaws, or venom that make us especially fearsome. What we do have is an exceptional mental capacity and the ability to adapt to a wide variety of environments. We are arguably one of the most successful types of multicellular organisms on earth, and in many ways, we can be considered the "fittest among the fit" precisely because of our adaptability and versatility. I admit, though, that this view is biased on the account that I am human. Sadly, it is undeniable that in a sense our "fitness" is causing the demise of many other fellow species on our planet.

From this point of view, the unit of selection in evolution is the *individual organism*. There are some differing opinions among scientists, however, about the specific level of selection. This includes opinions from both the scientists who actually work on evolution research and the proverbial "armchair quarterbacks," like most of us. Most likely, as with everything biology, there will be shades of gray. For our purposes, however, it suffices to view this phenomenon in light of the organism because, after all, organisms are the unit of reproduction. Populations do not reproduce in the strict sense, because a population grows by the combined reproductive rate of its members. Only individuals can reproduce, therefore passing along their genetic endowment. In other words, only individual organisms are able to pass on their DNA, which is the essence of the process of the perpetuation of life.

The majestic tale of evolution has taught us that all life on earth is related. The best available evidence indicates that all known organisms share a common ancestor. This common ancestor is unknown and unknowable I'm afraid, since any traces of its existence have surely disappeared over time. That said, as in a detective story, there are clues that allow us to paint a composite picture including many of the characteristics that this first organism could have possessed. This common ancestor has a name, LUCA (last universal common ancestor; more on this in the next chapter)[13] and is proposed to have lived about 3.8 billion years ago based on the best molecular evidence. We have many pieces of evidence that indicate this common ancestry and therefore the interrelatedness of life on earth. It is only recently that we have developed the tools that help us establish the connection between all forms of life that we see all around us. The best evidence that we have in this respect comes from the study of life at its most fundamental level. This level is the realm of cellular and molecular biology. We'll talk about it next.

> *The element of chance in basic research is overrated. Chance is a lady who smiles only upon those few who know how to make her smile.*
> —HANS SELYE

2

BIOMEDICAL RESEARCH

BIOCHEMISTRY AND CELL BIOLOGY LITE

TO BEGIN WITH, the very definition of "life" in the biological sense is elusive; this concept has proved notoriously difficult to characterize. One thing is for sure: life is more like a process than like a thing. We usually do not have any problems describing what life does, but we are most definitively at a loss when trying to state what life is.

Even though a universally accepted definition of life does not exist, we can list several traits that are common to all types of organisms we know of. We must keep in mind that these traits pertain to the concept of life "as we know it." After all, our type of life is the only example we know so far. There are many interesting speculations about alternate forms of life. However, at the time of this writing, there is no concrete evidence of such life forms. Therefore, from now on, whenever we speak of life, the "as we know it" phrase will be implicit unless otherwise noted.

Regarding the unity of life, we'll start by stating that every single type of organism described to date is built on chemicals; in turn, those chemicals are organized into a variety of molecules that can be classified into several main groups. These include proteins, lipids (a generic term for fats), carbohydrates (a generic term for sugars), nucleic acids and water, with a variety of other elements and compounds added for taste. In a very true sense, *life is chemistry*.

Structurally and functionally, the fundamental unit of life is the *cell*. A cell is the minimal ensemble of chemical compounds that, when organized in a very particular way, displays the characteristics of life (I know; it looks like a circular definition, but read on). In other words, a living organism is composed of at least one cell. Cells are entities separate from the environment and from each other by flexible, mainly lipid-formed structures called membranes. In addition to cell membranes, plants and certain microorganisms possess an additional boundary, a more or less rigid cell wall. The internal structure of a cell is anything but simple. There is a series of structures that will be present in most types of cells, depending on their complexity. Cells are classified in two main types, eukaryotes and prokaryotes. We'll see them in more detail later, but their main distinguishing difference is that the former contains a series of membrane-bound organelles (little organs) like the well-known nucleus, mitochondria, and so forth, which prokaryotes generally lack. The internal environment of a cell, where many of its essential chemical reactions take place, is called the protoplasm or the cytoplasm. These two terms mean slightly different things, but for our purposes, we can consider them one and the same.

For anything to be considered "alive," it must be distinct and separate from the environment. This "in" versus "out" separation is one of the main functions of membranes. A second essential function of membranes is as gateways through which information exchange takes place, as we will see in a moment. Even though membranes are mainly composed of lipids, there are other types of molecules located within them. These molecules are usually proteins of varying sizes, which act as receptors that recognize substances in the environment and deem them as beneficial or harmful, as well as for the purposes of communication within the cell itself or to communicate with other cells. Sometimes these receptors also control the transport of substances in and out of the cell. Other substances commonly found in membranes are carbohydrates, which act as labels that can be specific to different cell types and as energy-storing molecules.

Even though a living organism must maintain a barrier between its interior and exterior, an essential characteristic of life is that a living entity is an *open system*. An open system is capable of exchanging information (in the form of energy or chemicals) with its environment. In other words, a live organism must be capable of capturing energy and using it to maintain its internal order. Without this information exchange, there is no life. In this light, one of the essential characteristics of life is that it is capable of capturing energy from the environment and transforming it into a useful form. When an organism reacts to the environment by capturing energy, the main results are self-maintenance and growth. Some organisms are capable of synthesizing nourishing molecules by capturing energy directly from light (like plants) or even capturing energy directly from chemical reactions or similar means. Such organisms are collectively called *autotrophs*. Other organisms must consume other organisms to meet their energy requirements; we call these organisms *heterotrophs*. We humans are of course, heterotrophs.

When a living being reacts to the environment in any way, we describe this as behavior. All organisms display behavior, which allows them to seek out nourishment, as described previously, or to avoid interactions with toxic substances, since contact with these substances may prove harmful or even fatal. We usually associate behavior with animals, but it is important to point out that all organisms display this property. For example, unicellular organisms are capable of sensing their chemical environment so that they can swim toward nutrients or swim away from toxic conditions. Plants display behavior too, as we will see shortly.

All forms of life, without a single exception, use the energy stored in a molecule called adenosine triphosphate (ATP). ATP is not the only bioenergetically relevant molecule, but it is the one every single class of organisms studied so far uses—again, no exceptions to date.

One of the main characteristics of life is that it reacts to environmental conditions; that's how survival works. A living organism must survive to reproduce (which is the name of the game in evolution). Therefore, all organisms must be capable of fleeing or defending themselves from predators or becoming predators themselves, and sometimes both: an organism can be a predator to another form of life and, at the same time, a prey to yet another type of organism.

Finally, life has the capacity to reproduce. This means that organisms can generate (not quite identical) copies of themselves by the transmission and recombination of their genetic information. This genetic information in all kinds of life is invariably coded in a type of nucleic acid called DNA. In fact, in a very real sense, there is only one kind of life on earth, DNA-based life.

The story of the discovery of the structure of DNA and related macromolecules is just over 60 years old. We all have heard about it in one way or another. Here are just some of the main points. Nucleic acids like DNA and its close analog, ribonucleic acid (RNA), are *polymers*. A polymer is a chemical entity formed by smaller subunits, called *monomers*, often organized very much like the links in a chain. The individual links represent the monomers and the chain itself represents the polymer. In general, whenever we talk about polymers, we imply big molecules, usually referred to as *macromolecules*. In DNA and RNA, the monomers that form these nucleic acids are called *nucleotides*, which are molecules composed of three main parts, a specific sugar molecule (deoxyribose and ribose in DNA and RNA, respectively; that's where the names come from), which associates with a phosphate group and one of four different nitrogen bases. These bases in DNA are adenine (A), thymine (T), guanine (G), and cytosine (C). In RNA, the bases are A, G, C, and uracil (U). In other words, the language of life as we know it has an alphabet of four letters in DNA (ATGC) and four letters in RNA (AUGC). The rules of the language of life in its present form dictate that the words must be composed of three letters.

In itself, the information encrypted in DNA does not do anything in the sense that it is "trapped" within the molecule. In the way life works right now, the information in DNA is copied and passed on to RNA, which provides the point of exit of the DNA-coded information into a third type of macromolecule, proteins, of which we will talk about soon.

This sequence of events happens in all kinds of living organisms in our planet, again, without exception. For this reason, this is sometimes termed the "central dogma" of molecular biology. This central dogma is based on the fact that in itself, the information encrypted in DNA is not capable of directing the synthesis of proteins. Briefly (and *very* simplified), the path of genetic information starts with DNA, which is then copied into a specific RNA type by a process called *transcription,* and finally proteins are made based on the instructions encoded in RNA in a process called *translation*. RNA encodes the instructions to make proteins in three-lettered words called *codons*.

Proteins are probably the most versatile molecules within a living cell; they can form literally thousands of different structures that perform or control all the physiological processes that generate and sustain life.

Proteins are also polymers. The monomers that make proteins are called *amino acids* (Figure 2.1). There are hundreds of amino acids found in nature. Remarkably, life, from the smallest bacteria all the way to the blue whale, uses a set of only twenty or so amino acids, which are specified by the genetic code. This is yet another example of the unity of life.

The trick to making all these different proteins specific to an organism is the type of amino acids and the precise order in which they are associated with each other. Moreover, most amino acids are *chiral*, which means they can appear in two basic versions that are mirror images of each other. Life overwhelmingly uses one type over the other. The "R" group is the variable part of an amino acid; it is what determines its specific chemical properties, for example, charged versus noncharged, polar versus nonpolar, hydrophobic versus hydrophilic, and so on. Chemically speaking, proteins are made when amino acids react with other amino acids to form a special kind of bond called the *peptide bond*, as shown in Figure 2.1.

The type of amino acids in a protein largely determines the protein final structure and thus its function. However, the order of the amino acids is also

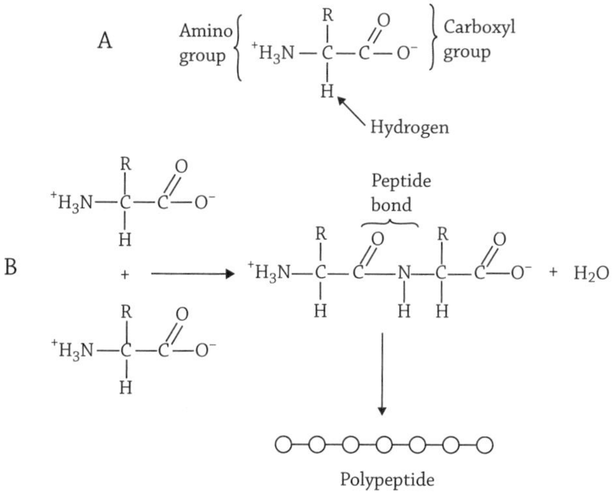

FIGURE 2.1. A. Typical amino acid structure. B. Formation of a peptide bond and polypeptide. Illustration by O. R. Pagán.

important. In other words, a short protein formed by amino acids 1-2-3-4 is not the same as another protein containing the same amino acids ordered as 4-1-3-2, or any of all the other possible combinations. This is very important especially if we consider that some proteins are composed of thousands of amino acids. If we think about all the possible combinations in which the amino acids in a protein that is 1,000 amino acids long can be arranged, we will get an idea of the many different kinds of possible protein structures that can possibly be generated. Moreover, a protein composed of 1,000 amino acids is not particularly big. The current record holder seems to be *titin*, a protein found in several versions associated with muscle tissues. An average titin molecule is close to 30,000 amino acids long! A point of nomenclature: when a protein is made by up to 30 amino acids or so, it is usually called a *polypeptide* or *peptide* for short.

There are some other entities in our life that display lifelike properties but that are not quite alive. We collectively call them *viruses*. Viruses seem to lie on the "borderline" between life and nonlife, as they are indistinguishable from inert matter, but they "come to life" when they interact with a specific host cell. In other words, they require a living cell to display lifelike properties, including reproduction. Some viruses use DNA as their genetic material, while others use RNA for that purpose. Interestingly, viruses are capable of evolving and adapting to their hosts; their genetic material mutates as in any living organism, providing them with the advantages of genetic variability. Thus, arguably, viruses are some of the most successful replicating entities known, but this story is beyond the scope of this book. If we do not count viruses as living organisms, it is safe to state that in any form of cell-based life, the genetic information is invariably stored in DNA.

HOW WE CLASSIFY LIFE

One of the strategies that humans use to try to understand the world, especially the living world, is to classify things in a more or less logical way. It is important to point out, though, that this logic changes over time as we obtain more information about the life forms in question. During the course of history, there have been various classification schemes that tried to impart some order on the living world that is all around us. In our slightly more scientific times, we call the branch of biology that dedicates itself to this effort *taxonomy*, not to be confused with *systematics*. Taxonomy is mainly about classifying known (and sometimes hypothetical) organisms into distinct groups. Systematics is the study of biological diversity in its evolutionary context. There are multiple criteria for organizing living beings into specific categories, but the first and most common criterion to try and do this is the description of the external appearance of an organism. In other cases, the way of life of a being was used as a classification criterion, for example, whether it lived on land or in the water, and so forth.

Remarkably, even before the genetics and molecular biology revolutions, taxonomists were very successful at figuring out the interrelationships of the many different organisms on earth. This is especially significant since they based their classification schemes mainly on structural similarities and differences between various forms of life—no physiology or cell or molecular biology involved. More recently, the availability and power of genetic information has confirmed and even extended many of these classification schemes, often in unexpected, interesting ways.

As science progressed, additional tools became available to describe an organism's traits, including its physiology and other characteristics that were not evident to the naked eye, for example. Later on, advances in genetics and molecular biology allowed us to compare different organisms by comparing their genes and gene products. In that way, we are able to ascertain the relatedness of various forms of life at their fundamental, molecular level.

As with many classification schemes, the taxonomy of organisms is a fluid science and therefore it is always subject to revision based on the latest available information as described earlier. The current classification scheme is based on a combination of morphological, physiological, and molecular traits, by which we are able to classify life on earth into three main *Domains* (Figure 2.2). Two of these domains include organisms called *prokaryotes*, which are relatively simple unicellular creatures whose major distinguishing feature is the absence of an organized, membrane-bound, DNA-containing nucleus. The two domains that include prokaryotic organisms are the *Bacteria* and the *Archaea*.

We are all familiar with bacteria. We generally know them as disease-causing (pathogenic) agents, but relatively few types of bacteria are pathogenic—although truth be told, some of those disease-causing bacteria are real troublemakers indeed. However, the vast majority of bacteria are perfectly harmless, beneficial even, as some bacteria are truly essential for our own survival.[1]

The second prokaryotic group, the *Archaea*, was only recognized as an independent domain of life in the late twentieth century.[2] Many archaea are *extremophiles*, which are organisms that thrive in extreme environments in terms of temperature, pressure, and other conditions. This is, of course, a somewhat biased classification, as we deem the conditions at which these beings live as "extreme" based on what we consider "normal" conditions from our perspective. Interestingly, even though the archaea look very similar to bacteria, at the molecular level, they are rather different, which was in no small part one of the reasons to warrant the creation of an entire taxonomical domain just for them. In fact, the archaea are in some ways taxonomically closer to a third domain, the *Eukarya*, which includes all organisms, unicellular or multicellular, that possess a DNA-containing cell nucleus. We humans belong in this category. Recent analysis of microbial life surveys suggest that a new class of viruses may warrant creating a fourth domain classification, but in my opinion, it is too early to tell if this will hold.

The classification step right under the Domain category is the *Kingdom*. There are six of them; the first two correspond to the Bacteria and the Archaea mentioned previously. The other four are composed of eukaryotic organisms, including *Plants, Fungi, Animals,* and *Protists*. Of these four kingdoms, only the protists are exclusively unicellular (well, most of the time anyway).[3] We, of course, are part of the animal kingdom.

The *binomial system of nomenclature* is a useful, established taxonomical method, which assigns a *scientific name* that consists of a proper name (genus) and a last name (species) unique to each kind of living thing. The naturalist that came up with this system was Carl von Linné (you may know him by his Latinized name, Linnaeus) in the late 1700s. In this scheme the scientific name for humans is *Homo sapiens*. Scientific names are customarily shown in *italics*, with the genus capitalized and the species in lowercase, and are also usually descriptive of a particular set of characteristics of the organism (like humans, *Homo sapiens* = "Wise human"). Sometimes a particular type of organism gets its scientific name based on who discovered it or on the set of circumstances that led to its discovery.

My first exposure (and maybe yours too) to the binomial system of nomenclature was by watching the old Looney Tunes' Roadrunner and Wiley E. Coyote

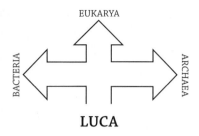

FIGURE 2.2. Life domains. LUCA: Last Universal Common Ancestor. Illustration by O. R. Pagán.

cartoons, shown for the first time in the late 1940s (I used to watch them in the early 1970s though!). A frequent feature of each episode was to show a kind of mock nature documentary that gave false Latinized scientific names for the protagonists, for example, *Eatibus anythingus* for the Coyote and *Velocitus tremendus* for the Roadrunner among many other made-up scientific names.

There are many interesting examples of curious reasons to assign scientific names to a newly discovered species; you may be familiar with one of my favorites, the story of the coelacanth. The coelacanth is a type of fish presumed extinct for as long as the dinosaurs themselves went away, about 65 million years ago. In the late 1930s, a museum curator in Cape Town, South Africa, Marjorie Courtney-Latimer, discovered a coelacanth specimen in a fisherman's catch. Since then many other coelacanth specimens at various geographical locations have been observed and even a distinct separate species has been identified. To make the long story short, the original coelacanth's scientific name is *Latimeria chalumnae*. The genus (*Latimeria*) honors Courtney-Latimer as the discoverer and the species (*chalumnae*) comes from the word *calumny* (meaning "false charge"), essentially indicating that, to paraphrase Mark Twain, the rumors of the coelacanth's demise were greatly exaggerated.

Between the kingdom and the scientific name there are other classifications. As we go "down" in the classification scheme, we find fewer and fewer creatures per taxonomic unit until we get to the species, of which there is only one (well, more or less; we can talk about subspecies in some cases). If you wish to know more about biological taxonomy, there are some excellent books out there that provide accurate, up-to-date information alongside the historical context of the development of taxonomy.[4,5,6]

ANIMAL MODELS AND BIOMEDICAL RESEARCH

Why are nonhuman organisms useful in biomedical research? Animal models have proven once and again to be very useful to understand aspects of human physiology. This is possible simply because in essence, there is one, and only one, type of life in this planet, life that is carbon based and uses DNA as the genetic material.

It is a well-established fact that all life uses the same basic building blocks as described previously; also, there are many commonalities in biochemistry and physiology (by the way, physiology is nothing other than applied biochemistry, but don't tell anyone, especially the physiologists). All these similarities are very important because they mean that the biology of "lower" organisms is a pretty good approximation to human biology (or is it the other way around?).[7] This is why many of the chemicals that we use to eliminate pests, for example, can be harmful to humans as well and why molecules evolved in, say, venomous snakes and spiders to allow them to hunt for food are capable of hurting us too. Think about this: we are not the intended prey of a black widow spider or a rattlesnake, for example, yet we are nonetheless affected by their venom, oftentimes with lethal results.

From this point of view, biomedical research using nonhuman animal models provides scientists with useful insights on disease conditions and on the development of pharmacological agents. That said, animal research is a controversial topic to say the least; however, it is undeniable that such research has advanced biomedical sciences in ways that would have been much more difficult or even impossible without it. Many people today are alive because of research performed on animal models; this is a fact, end of story.

It is important to keep in mind that we can study nature for its own sake, without any explicit expectation of finding an application for that research. In fact, this was the main motivation for most of the naturalists we referred to earlier in chapter 1. To be completely honest, even though I really think that modern research is incredibly exciting, especially in light of the potential benefits for humanity, there is something to be said about simpler times, when a person was able just to follow his (as they were usually men, as unfair as that was) curiosity and admiration for nature. Maybe if and when we conquer all diseases (a guy can dream), we scientists can return to our ancestral roots and learn about nature for its own sake. Then we can join the wonderful and capable group of the so-called amateur naturalists in the common ground of our shared appreciation of the universe.

Where was I? Right, research applications. This is not to say that there will be no such potential applications of research done for its own sake—far from it. There are many examples of unforeseen applications of fundamental research, as we will see later. In biomedical research, on the other hand, the explicit motivation for a given research activity is to obtain possible practical applications.

Probably the best-known aspect of experimental biology is the so-called *in vivo* ("truly alive") experiments, in which a complete organism is studied. Again, the true significance of a molecular process can only be visualized in the whole organism.

There is a series of organisms that have traditionally been used in biomedical research. Here we will find the usual suspects, like rats and mice, which are well-characterized vertebrate models. Also, we are all familiar with probably the most controversial aspect of research using animals, namely, research on primates. A good rule of thumb is that the closer an organism is to humans in evolutionary terms, the more similar its physiology will be to ours. However, in principle, any type of organism can be used in biomedical research. That said, a fair question would be: why are some but not all kinds of organisms used in such research? There are several reasons for this, most of them related to practical considerations. The first thing that a biomedical scientist implicitly assumes about a model organism is that the aspect of biology that will be studied will be reasonably similar to the same process in humans. This would be true of virtually any type of being in this earth of ours; of course, some will be closer than others.

A second consideration is the practical advantages of using a given species. Let's take the study of the fundamental principles of learning and behavior for example. In principle, it would be possible to use, say, whales or tigers for such purpose, but it would not be particularly practical, would surely not be cheap, and would certainly be very dangerous. So, to study the fundamentals of behavior, whales or tigers would not be a good choice. Simpler, smaller, and less dangerous animals will be better suited for such purposes. Of course, it is an entirely different matter if you want to study *specifically* the behavior of tigers or whales. Then there is no other choice and scientists will make this work.

A third criterion for choosing a research species is a special characteristic of a given species that may facilitate the experimental process. Let's see some examples.

One of the best-known research subjects in biology is the fruit fly (*Drosophila melanogaster*), which was and still is one of the ideal organisms to study heredity due to several key properties including its well-defined genetics, relatively fast life cycle (about 14 days), and ease of culture. These invertebrates are also very useful and convenient to study behavior and as an animal model for drug abuse research.

The zebrafish (*Danio rerio*) is an established animal model in molecular and developmental biology and is rapidly becoming a popular model in neurobiology as well. This organism also has well-characterized genetics and behavior as well as a relatively short life cycle (about 12 weeks). Again, the relatively easy culture and keeping methods play an important factor in this animal choice. An additional advantage of the zebrafish is that it is a vertebrate; its biology is "closer" to ours compared to the biology of invertebrates, for example.

That said, being an invertebrate is not a deal breaker in biomedical research. In some cases, simplicity is the key. A type of roundworm (*Caenorhabditis elegans*) has proven invaluable to elucidate physiological processes common to many forms of life. In addition to its known genetics, ease of culture, and short life cycle (about 4 days!), this organism has exactly 959 cells, no more, no less, excluding its sex cells (this is under normal circumstances). Of the total number of cells in a normal *C. elegans*, exactly 302 are neurons (again, normally). Since the mapping of all cells in this organism is complete, this makes it possible for a scientist to literally eliminate specific cells, one by one even, and see what happens!

Sometimes the presence of an unusual structural feature on a type of organism is what makes it an ideal subject for a specific purpose. In neurobiology, one of the early stars was the squid, with its giant nerve cell (not to be confused with the giant squid, which I am quite sure has giant nerve cells as well). These giant cells are big enough to see with a simple dissecting microscope. Also, their size makes them easy to manipulate, which facilitates experimenting with them. Many of the fundamental physiological properties of the nervous system were first characterized in this model.

These are just some of the examples of the many types of creatures that represent useful biomedical models. Even organisms other than animals have proven very useful research-wise. Other surprisingly suitable types of research subjects include plants like the mouse ear cress (*Arabidopsis thaliana*) and a type of yeast (*Saccharomyces cerevisiae*). These two organisms are contributing to the understanding of fundamental cellular properties and important mechanisms of development. These are just a few examples of some of the best-known non-human model organisms currently used. There are many more.[8]

Let's now talk a little bit about other strategies used in biological research in general, but before that, it is important to understand that the terms *biological* and *biomedical* research can be used interchangeably, but in strict terms, biomedical research implies a goal-oriented effort, like curing or alleviating a disease, for example.

Biological research can be done at different levels. As we just saw, when using intact, living organisms, we call it the *in vivo* approach. Going back to basics, the most fundamental level is the study of biochemistry. Keep in mind that the phenomenon of life seems to arise from the interaction of multiple components. This means that molecules like proteins, nucleic acids, and so forth are not alive in and of themselves. However, in many cases, we can use *cell-free* preparations, which contain the molecules of interest. These molecules retain their chemical and physiological properties (albeit for a limited time), allowing the researcher to study them. This particular approach is often called *in vitro* ("in glass," as in a glass test tube).

The next level is, of course, cells. Here you can study cells by themselves, taken directly from tissues or even cultured, for example. In many cases, these cells can be kept for a long time in culture, which allows a more detailed exploration of their properties. Furthermore, sometimes you can even modify these cells by incorporating genes not usually found in them. This cell-based approach is called *ex vivo* (loosely translated as "alive, but not in their natural state"). Another type of *ex vivo* experiment is when only a part of an organism is studied, like isolated muscle fibers, for example.

A very interesting approach is called *in silico* ("in sand"), which refers to the use of computers to model biological processes (the term *silico* also refers to the origin of the Silicon Valley region's name). Now, this is a trickier method, because as useful as computer modeling can be, it is only as good as the information available about a system. An early computer science adage is GiGo (garbage in, garbage out), meaning that if you give the computer faulty information, you will get faulty results. In this sense, we know quite a bit about the organization of certain biological systems; in other cases, not so much. In these latter cases, the *in silico* approach must be taken with a grain of... salt.

As important as animal research is, it is far from perfect. For example, there are inherent limitations when using specially bred laboratory animals like mice and rats. In many cases, these breeds are genetically homogeneous, meaning they display lesser genetic variability when compared to a natural population. Please recall that genetic variability plays an essential role in biology, particularly in the survival of species. When a population of organisms is extremely similar in terms of its gene pool, this is not generally good news. First, this is not a natural occurrence, as this degree of genetic similarity does not usually apply to a population of organisms in the wild. In fact, genetic variability can be used as an indication of the "health" of a species. If we want to see a very close example of genetic variability, we can always look at ourselves, at human populations. Just look around wherever there are people. With the exception of identical twins (and even then this is not "absolutely absolute"), there is a fair degree of genetic variability in humans.

Please allow me a brief digression to introduce a couple of terms: the genetic endowment of an organism is called the *genotype*. This is essentially its biological information. When this biological information is expressed, it interacts with the environment to generate something called the *phenotype*. The phenotype is thus the expression of this genetic information in light of the external world. The phenotype is usually associated with appearance (blue eyes, brown hair, etc.) but also includes many other characteristics, like how one reacts to medications or how sensitive one is to certain smells, for example.

Back to genetic variability.

When a population is genetically diverse, it has a better chance to withstand changes in its living conditions. For example, if a new virus appears, the more genetic diversity there is in a given population, the better the chance that resistance genes will be present in some individuals. These genes will improve the survival chances of those individuals, which will pass on those resistance genes to their offspring. It is simply a numbers game. The opposite happens with lesser variability. In a more homogeneous population, there is a lesser probability of finding favorable genes for all possible eventualities.

A possible reason for less genetic variability in a population is inbreeding, which, in addition to diminished genetic variety, increases the chances of expressing lethal genes. A specific example of this is the African cheetah, the fastest land animal. These big cats are endangered, which is bad enough in itself, but the lack of normal genetic variability between individuals in their population is also contributing to their decline. We can visualize this scenario with an example from humans. Our immune system is in charge of detecting and repelling "molecular invaders," meaning distinguishing self from nonself. This is why organ and skin transplantation in clinical medicine is so problematic; even closely related individuals may not be a match to each other, due to differences in their individual genetic makeup. For example, if a person needs a kidney donor, sometimes even siblings may not be a perfect match. In contrast, any modern cheetah would be a perfect donor to another one, with no rejection at all, even if they come from a different population; they are that similar.

All the concepts discussed previously have implications on the use of dedicated laboratory strains of rats and mice for biomedical research. In many cases, these strains are genetically homogeneous, which will skew any observations because of their lack of genetic variation. These laboratory breeds are perfectly fine when studying the "normal" physiology of an organism, which by definition will include the most fundamental properties absolutely required for life, which will be shared by all organisms in a species. For example, let's take the famous Krebs cycle: In this cycle, most of the precursors necessary to eventually synthesize adenosine triphosphate (ATP, the energy currency of life) are generated. This basic physiology will be more sensitive to mutations, meaning that mutations in the genes that produce proteins related to the Krebs cycle will most likely make the process less efficient. Therefore all individuals of a particular species, just by their virtue of being alive, must have virtually identical inner workings regarding fundamental processes.

Genetic homogeneity becomes a problem when we want to study aspects of physiology that are not necessarily critical for life, for example, how a human will react to alcohol. Some people will naturally possess genes that allow them to be able to metabolize alcohol amounts that will be toxic to another person with different genes. However, if that person who is more sensitive to alcohol intoxication avoids drinking, those genes will not effect that person's survival.

Similarly, it is entirely possible that just by chance a specific mouse or rat breed has genes that render it more or less sensitive to the exposure of a particular chemical agent or drug candidate. Thus, drawing conclusions about any pharmacological data obtained using such a breed will potentially not apply to a "normally" diverse rodent population, let alone humans.

Moreover, there are some cases that show that when rodent models have been used to assess the safety of experimental medications, no indication of any negative effects were observed, but when the drugs were used by humans, they were proven to be harmful. A famous case was the drug thalidomide, developed in the 1960s as a medication against morning sickness in pregnant women. This medication was soon found to have serious developmental side effects. The history of thalidomide has additional aspects other than the overinterpretation of animal studies data; nonetheless, it represents a cautionary tale against solely relying on animal models to reach conclusions about a particular drug's usefulness, efficacy, toxicity, or overall safety.

Alternate animal models include primates. These are the organisms most closely related to us in an evolutionary sense. However, even chimpanzees, which, with gorillas, are our closest extant (living, as opposed to extinct) relatives, display significant, important physiological differences from us. For example, pathological conditions like cancer or human immunodeficiency virus (HIV) infection seem to be expressed differently in chimpanzees and in humans. Also, there are many ethical concerns that need to be taken into account when considering doing research on primates. Do we have the right to experiment on them, by virtue of us being a more "advanced" species? This is only one of the many ethical questions that are relevant when we think about research using animals. That said, I hope to provide some arguments in favor of using alternate animal models, preferably models that display their natural, healthy genetic variability, in biomedical research.

It is important to point out that as powerful and important as animal models are in biomedical research, ultimately, human trials are required to approve the use of any drug in human medicine. There is simply no better way to assess the drug's effects.

BASIC OR FUNDAMENTAL RESEARCH?

A very well-known maxim states that "it's not what you say, but how you say it." Well, that is true, but I'd like to point out that the opposite is also true; in many cases what matters is actually *what you say*, independently of how you say it. The meaning, or rather the perception of the meaning, of a term is a critical factor when using words to communicate with each other. The way you may perceive a particular word as applied to a certain topic may affect your opinion of it, for better or worse. Words are important!

In science, particularly in scientific research, we have a sadly perfect example of one of the misconceptions that can arise from the inaccurate perception of a word. I am talking about the phrases *basic science* and *fundamental science*. I must disclose that this is one of my scientific pet peeves.

According to the dictionary, the definition of *basic* is "of, relating to, or forming a base; fundamental."[9] In turn, using the same source, the definition of *fundamental* is "Forming or serving as an essential component of a system or structure; central."[10]

In other words, this means that the terms *basic* and *fundamental* are therefore one and the same. The thing is that in everyday language, in some cases, the term *basic* implies simplicity, as in something that is less complex. In science, this is not so. An unfortunate erroneous implication of this take on the word *basic* is that basic science is of a lower level or even less important or relevant than applied or medical science. This could not be further from the truth. In a very real sense, applied or medical research would not be at all possible without basic (fundamental) research.

Think about this: an engineer could not even begin to design, say, a car, a building, an airplane, or an electronic device without knowing her physics and mathematics. Similarly, a pharmaceutical or medical scientist will not be able to discover the next wonder drug without taking into consideration human physiology. Basic science and, by extension, basic research are indispensable components of the scientific endeavor. Personally, I would rather talk about *fundamental* as opposed to *basic* research precisely to avoid this misconception.

One of the main advantages of the proper practice of science in general, not just biology, is that it is self-correcting. This means that when scientists report original research results, we are supposed to disclose the used materials, methods, equipment, and so forth as specifically and as detailed as possible. The idea is that in principle, the results should be reproducible, as reproducibility goes (meaning within reason). It is important to realize that this is especially critical in biology. Living systems are so complex and sensitive to the environmental conditions that identical results are seldom, if ever, seen. In fact, consistently identical results in biology are actually rather suspicious. The idea is that if I report some results from my laboratory, another research group anywhere in the world should be able to replicate these findings at least in qualitative (non-numerical) terms, provided that the materials and methods are reasonably similar. Ideally, quantitative (numerical) reproducibility is desired, but this does not mean that quantitative results should be identical; rather, this means that the two sets of results are not different in statistical terms. This is a story for some other time though.

Another aspect of the self-correcting nature of science research is peer review. When you submit your research for publication, a journal editor sends your submission to several scientists knowledgeable in the topic. Ideally, they will read it critically and may suggest modifications and even further experiments. In some cases, they may reject the paper altogether. Ironically, in the vernacular, peer review is not an exact science, simply because scientists are humans. In general, the system works really well considering, but the human factor is ever present. Scientists may show bias, envy, and even good old stubbornness, depending on the nature of the report. In general, this is avoided by sending the paper to more than one scientist. The system is not perfect and it has been heavily criticized, but, so far, an adequate alternative has not been invented yet.

Another important aspect of fundamental research is that in most cases, it is next to impossible to predict the practical applications of "pure" science as opposed to applied science. In principle, any scientific discovery has the potential for practical applications. We could talk about many examples, like the development of X-ray imaging and magnetic resonance imaging (MRI) technology, both of which have proven to be invaluable tools in diagnostic medicine. These techniques are a direct outcome of physics and physical chemistry research, respectively. This is a perfect illustration that shows that discoveries in one field often find their applications in another field.

Other examples that show how unpredictable the practical applications of fundamental research can be include the development of medicines derived from nonhuman organisms. For example, there is a group of drugs called statins that have proven very useful in treating patients showing higher-than-normal levels of blood cholesterol, which may have implications for cardiovascular diseases.

Originally, these drugs were discovered in fungi. Who in their right mind would have thought to check out molds in search for the treatment of cardiovascular conditions? This is just one of the many documented cases where naturally occurring compounds are the source of useful medications. I feel very strongly about this. In fact, this was the first line of my PhD dissertation:

> Nature is the best chemist. During the course of evolution, through literally millions of years, a wide variety of organisms have developed substances used for defense against predators, or to become predators themselves. As part of the evolutionary process, chemical structures beneficial for the survival of the organism are conserved; many of these molecules include small organic toxins.[11]

A particularly relevant quote that nicely complements my thoughts comes from one of my favorite scientists, Edward O. Wilson:

> Revolutionary new drugs have rarely been developed by the pure insights of molecular and cellular biology.... Rather, the pathway of discovery has usually been the reverse: the presence of the drug is first detected in whole organisms, and the nature of its activity is subsequently tracked down to the molecular and cellular levels. Then the basic research begins.[12]

Toxins and venoms are some of the most fascinating substances that can be found in many types of living organisms. In general, toxin and venom components are used as the means for an organism to defend itself or to kill other organisms for nourishment. Even though the terms *toxin* and *venom* are often used interchangeably, there are important differences between the two. Toxins are usually simpler organic compounds that are not stored in specialized organs and require the "cooperation" of the intended target. For example, plants produce many types of defensive compounds (more on this in the next chapter), but to induce their effect they must be actively consumed by their target (insects, grazing animals, etc.). On the other hand, venoms consist of a complex combination of many different toxic substances (many of them toxins in their own right), which can be small organic molecules, peptides, or even large proteins. In general, a venomous organism possesses anatomical structures to store the venom and a mechanism to deliver it, usually in injectable form, such as fangs, as in some snakes, spiders, and so forth.

Some of the most interesting venomous organisms are certain types of marine snails, the cone snails. This is a relatively recent class of snails, as they seem to have evolved about 50 million years ago. They are widely distributed in marine environments, with species found in every major ocean. Many species of these mollusks are brightly colored and are therefore highly prized by serious shell collectors. All species of cone snails are predators. Some cone snails hunt marine worms, several species hunt other snails, and yet some other species feed exclusively on fish—yes, fish. And by the way, did I mention that cone snails are, well, snails? Snails are invariably S-L-O-W. Cone snails are no exception. How are these organisms capable of hunting fish?

Cone snails have evolved a series of very potent venoms, with toxin components targeting several aspects of neurotransmission, namely, various types of ion channels (more on them in the next chapter). These toxins produced by cone snails are known as conotoxins. Conotoxins are short peptides, usually no longer than 20 amino acids long. Cone snails display several strategies to capture their prey. Sometimes, when a cone snail is hunting, it extends a proboscis, which has at the tip a harpoonlike structure through which the venom is injected when the prey is stung.

When a fish is stung, several types of conotoxins present in the venom simultaneously block various types of ion channels important for the proper function of the fish's nervous system. This happens very fast; the fish is paralyzed within two seconds or so! If you think about it, that venoms work quickly makes perfect sense, since if the venom were not that fast acting, the fish would be able to swim away before becoming paralyzed, and by the time the snail got to the fish, it would have probably been eaten already by something else and therefore the poor cone snail would go hungry.

The story of how conotoxins were discovered starts in the 1700s, when the first reports of fatalities associated with cone snail handling were described. The dangerous nature of many species of cone snails was very well known, albeit in an anecdotal way. In the late 1950s, the first reports of fish-hunting snails were published, and in the 1970s, there were early attempts to isolate and characterize some of their venom's components. The "modern" era of conotoxin research began with the work of Dr. Baldomero Olivera, of the University of Utah. Many different types of conotoxins were isolated and characterized in his laboratory. Since then, other research groups in Australia, Europe, and the Americas have joined the hunt to help study this promising class of pharmacological agents. Incidentally, the history of conotoxin research is a wonderful example of how undergraduate students can make significant contributions to scientific research.[13]

In at least one case, a cone snail toxin has already proven useful in human medicine. Ziconotide (Prialt) is the synthetic version of a conotoxin that blocks a particular ion channel important in neurotransmission. Ziconotide is approved for patient use and is currently given to cancer patients suffering with otherwise intractable pain. This example is a very direct illustration of the importance of biological research. Who would have thought that by studying a marine snail you could alleviate the suffering of a cancer patient?

Ziconotide is only one example of a cone snail compound. Some other conotoxins seem to have cardio- and neuroprotective properties. Scientists think that there are about 700 species of cone snails, each of those with up to 200 different venom components. This translates to about 140,000 possible compounds! How many cone snail compounds with useful properties are waiting to be discovered?

By extension, think about all the many different types of life on this planet. Where will the next wonder drug come from? From which organism will the next breakthrough with potential applications to human welfare be discovered? The conservation of our biodiversity is not only a matter of science, aesthetics, or morality but also makes medical and economic sense.

I would like to finish this chapter by restating that absolutely every aspect of biomedical research is related to evolutionary concepts. This is remarkable in itself, which reminds me of the words of Aristotle, "Philosophy starts in wonder and wonderment."[14] In my mind, one of the scientific disciplines that best represents Aristotle's way of thinking is the subject of the next chapter. In there, we will explore one of the most interesting aspects of biology: neuroscience.

… II

The Science of the Brain

> *The chief function of the body is to carry the brain around.*
> —THOMAS A. EDISON

3

AN INTRODUCTION TO THE NEUROSCIENCES

NEUROSCIENCE OR NEUROBIOLOGY?

FOR ALL PRACTICAL purposes, we can consider neuroscience and neurobiology one and the same scientific discipline. However, strictly speaking, neurobiology is really about the actual biology of the nervous system, as if you were studying any other system, like the renal or circulatory systems, for example. On the other hand, the scope of neuroscience is *explicitly* interdisciplinary in the sense that it explores the nervous system from multiple perspectives, which include but are not limited to philosophy, mathematics, chemistry, psychology, and good old biology, among many others. That said, please remember what we said about biology in chapter 1. Any branch of biology, neurobiology included, is truly interdisciplinary whether we realize it or not. To truly understand biology, you need to know chemistry, physics, and so forth; so you see, we can safely consider neurobiology and neuroscience the same thing for our purposes. Nonetheless, the term *neuroscience* is perceived as more interdisciplinary and therefore I will use that term in this book.

We can make a very good argument to consider the neurosciences as maybe the very first biological science that humans practiced. Let's explore this idea in some more detail.

If the current ideas on how human culture developed are correct, at first (I almost said "In the beginning...," but that phrase was already taken), we were hunter/gatherers. Gathering is hard work and it takes brainpower to recognize edible plants, but at least they are not running away from you. However, from the point of view of hunters (humans or otherwise), it is a good idea to get to know their prey, and the most direct way to do that would be to understand aspects of animal behavior. There is evidence indicating that early humans took notice and remembered specific prey behaviors, including migratory patterns and diurnal or nocturnal activity, among others. When they used such information while hunting, they were in fact applying neuroscience principles! Of course, our ancestors were able to use this knowledge only because they could learn from experience themselves. Therefore, because of their behavior, generated by their own brains, our predecessors enhanced their chances of survival and eventually left offspring, which is, after all, the point of life in its biological sense.

NEURONS

Animals interact with their environment by receiving information through their nervous systems. The interpretation of this information triggers the expression of behaviors that may allow an organism to survive. The *neuron* (Figure 3.1) is the basic cellular unit of any nervous system. The word *neuron* itself was coined in 1891 by the German anatomist Wilhelm von Waldeyer-Hartz.

There are many types of neurons, but they all share several structural features that allow them to serve as communication entities. As do all cells, they have a main body with its associated organelles; additionally, they possess specialized structures called *dendrites*, whose main function is to receive signals from neighboring neurons. A typical neuron receives multiple inputs from dendrites. Once these inputs integrate within the neuron, the signal travels through a usually long structure called the *axon*, which connects to the next neuron's dendrites through connections called synapses, and so on (Figure 3.1). There are many variations of this general theme depending on the type of neuron; the specific details will depend on the actual function and location of the neuron within a given nervous system (Figure 3.2).

One of the most interesting characteristics of neurons is that they are remarkably similar between animal species. For example, when observing neuronal cells under the microscope, it often takes an expert to determine the type of organism that they belong to, as neurons are not structurally or functionally simpler in "lower" animals or more complex in more "advanced" animals. For example, under the microscope, a neuron from a worm is rather difficult to distinguish from a human neuron. This fact is even more remarkable when we realize that neurons from a worm and from a human, for example, came from quite different organisms separated by hundreds of millions of years over the course of

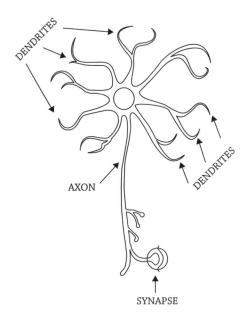

FIGURE 3.1. The main parts of a typical neuron.
Illustration by A. G. Pagán.

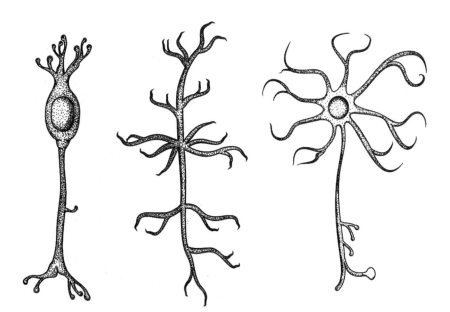

FIGURE 3.2. A sample of various neuronal shapes.
Illustrations by A. G. Pagán.

evolution. Another way of looking at this is to say that human neurons do not have any specific feature unique to them that makes them fundamentally different from other types of neurons.

In addition to their close structural similarities, all nerve cells, regardless of the animal they come from, share a common physiology; in fact, many of the fundamental properties of all nervous system cells were initially characterized using invertebrate animal models like squids, for example (see chapter 2). Thus, the study of many types of animal models has proved once and again their relevance to understanding the fundamental properties of nervous systems in general.

Essentially, the main characteristic of nervous systems is that groups of neurons communicate with each other to coordinate virtually all processes within the organism, including the generation of behavior. Currently, it is a well-established fact that nervous systems in general and brains in particular are composed of closely associated yet separate nerve cells. However, this fact only began to be apparent a little more than 150 years ago and was actually fully established only about a hundred years ago. Let's see a brief account of how this happened.

THE SPANIARD AND THE ITALIAN

As with many scientific discoveries, the history of how scientists established that neurons were the nervous system's building blocks is rich, intricate, interesting to say the least, and full of curious tidbits and anecdotes courtesy of human nature itself.

You see, after love, there are only three other really great things in life: *science, religion,* and *gossip* (well, four other really great things, if we were to include *history* in this list). The beauty about loving science is that with science, it is entirely possible to have it all!

Let's think about this last statement for a second. In most cases, whenever you want to see the whole picture of anything, to really understand any fact or observation in its proper context, you have to take into consideration its *historical background*. Science is no exception.

Also, science is done by humans (at least on this planet), and humans are, well, human, with everything that comes with the territory. This includes a healthy dose of *gossip* and *conflict*. It is undeniable that people gossip about the things that interest them; scientists are no different. Actually, Francis Crick (of DNA fame) advised beginning scientists to discover their true scientific calling by invoking the so-called gossip test.[1] More on this later.

Additionally, another characteristic of a scientific discovery is that whether we realize it or not, and furthermore, whether we like it or not, any such discoveries have at least the potential of changing our self-perception in the grand scheme of things and therefore our opinion of our true place in nature. In other words, scientific discoveries can alter our philosophical outlook on life. Moreover, scientific knowledge, for good, bad, or worse, frequently influences our *religious* and *spiritual tendencies* (or lack thereof) in one way or another.

This philosophical dimension is especially relevant if we want to understand the brain; after all, it seems that what makes us who we are is a direct consequence of the physical makeup and workings of our brain and nervous system.

The neuron story has many of the elements described previously and has been skillfully told in all its juicy and truly epic detail elsewhere.[2,3] However, I cannot fail to give you a brief account of it, as this is one of the most engaging chapters in the history of neurobiology, not only for its subject matter, but also because of its human main characters. These scientists had interesting personalities that undoubtedly influenced their outlook on life, and by extension their temperament and hence their scientific drive and behavior.

In the 1800s one of the most important questions people started to explore about nervous systems was whether they were composed of separate, individual cells, as every other organ system seemed to be, or whether they were a series of continuous interconnected compartments organized as a unit. Please keep in mind that this was the same century when Schleiden and Schwann articulated the *cell theory*, which, in brief, explicitly stated that the cell was the basic unit of life. In principle, a nervous system should be no different from any other organ system. What made nervous systems so distinct from other bodily systems?

One of the main points of contention when thinking about the brain in terms of the cell theory was that even when scientists observed nervous tissue using the best microscopy techniques available at the time, nervous tissue looked relatively nondescript. In many cases there was no undisputed, clear resolution of distinct structural features, and even in some cases where observation suggested structural features, those observations were often dismissed as artifacts.

Moreover, cells within the nervous system seemed to display a wide range of sizes and shapes (Figure 3.2), with no evident unifying features, which added to the confusion. Not surprisingly, the fine structure of nerve tissue was a source of intense debate, since it was evident that in virtually every other type of tissue in plants and animals, cells were clearly independent units that displayed more or less the same shapes.

Therefore, in contrast with the cell theory, which did not generate a great deal of controversy, the *neuron theory* was vocally and emphatically opposed by proponents of the alternate *reticularist theory*, which stated that neuronal tissue was a continuous network with no independent parts. The conflict between the schools of thought represented by the "neuronists" and the "reticularists," respectively, has been described as "fierce," and it was likely influenced by factors that were not strictly scientific, since some of these factors included discussions pertaining to the place of humans in the grand scheme of things, as we saw earlier. Another related factor contributing to these disputes was the struggle between the *reductionist* and *holistic* approaches to studying nature, which is a usual cause of spirited debate in the biological sciences (see chapter 1).

Some authors have pointed out that the neuron theory does not exactly represent a logical extension of the cell theory. In other words, both theories, although similar, are not strictly equivalent in the sense that the neuron theory goes beyond the cell theory in terms of its assertions about the structural features and function of the nervous system. However, for our purposes, it is safe to consider the neuron theory an extension of the cell theory, at least as a first approximation. In other words, we can safely say that the neuron theory complemented the cell theory.

Santiago Ramón y Cajal, from Spain, and Camillo Golgi, from Italy, were not the only characters in this chapter of the history of the neurosciences, but they are without a doubt the two undisputed protagonists. Golgi and Cajal[4] were the most prominent advocates for the reticularist and neuronist schools of thought, respectively. These two physicians/scientists shared a singular, complex bond, described by Cajal as "What a cruel irony of fate to pair, like Siamese twins joined back to back, (yet) scientific adversaries of such contrasting character!"[5]

Despite Cajal's opinion about the personality differences between Golgi and himself, with the advantage of historic hindsight we can confidently state that in fact it is pretty evident that they shared many scientific personality traits. Some of these qualities included a passionate dedication to their work and an intense fascination with and interest in the natural sciences in general and nervous systems in particular. Both were also true pioneers in cell biology research. Cajal has actually been described as one of the original neurobiologists, due to his masterful integration of various scientific disciplines and his outstanding scientific output. In fact, it has been argued that neuroscience itself as a discipline started with Cajal;[6] but then again, the same thing has been said about Golgi.[7]

As we saw previously, one of the arguments the reticularists used against the neuronists was that in most cases, the structure of nervous systems as observed under the microscope appeared rather fuzzy, without clear boundaries between independent cells. One tried-and-true strategy in microscopy practice to try to distinguish individual cell types was to stain tissue with chemical dyes to enhance contrast. Even though microscopists since Leeuwenhoek himself tried to stain nerve tissue to make it more visible, the 1800s was an era when this practice was in full swing. This was about the time when advances in organic chemistry contributed to the development of various synthetic stains and dyes, which were enthusiastically used by most microscopists. This was also the time when photographic techniques were being developed for the first time; these techniques also played an important (central even) role in this story.

In the 1870s Golgi and many others tried various combinations of dyes to try to visualize nerve tissue. Eventually, Golgi discovered and developed a staining technique that he called "the black reaction" (*la reazione nera*), using potassium dichromate, osmium tetroxide, and silver nitrate in specific proportions. The result of the black reaction was the precipitation of particles of silver-based compounds within neurons that appeared very dark against the bright field of view of the microscope; this rendered beautiful, high-contrast images. Oddly, only about 5% of the neurons were labeled (to this day, nobody really knows quite exactly why it is so); however, the stained neurons showed exquisite detail. Golgi published his results in a short paper in the journal *Gazzetta Medica Italiana, Lombardia*. Despite its evident usefulness, the use of the black reaction was not widespread, partly because the results were inconsistent and few of the cells were labeled, partly because Golgi published in an obscure journal, and partly because he published in Italian, not in any of the scientific languages of the time, which were mainly German, French, and, to a lesser extent, English.

Strangely enough, Golgi's early representations of nerve cells seemed to depict independent cells, and even in his later research years a member of Golgi's research group, Dr. Aldo Perroncito, obtained results quite similar to Cajal's. Therefore, it is unclear why Golgi held fast to the reticularist view. In all fairness, even when some inflexibility and stubbornness may have influenced his thinking, Golgi's support for the reticularist camp was not entirely without reason. Among other things, he argued that, based on his view of electrotonic (as opposed to chemical) transmission, direct contact between nervous fibers was not even necessary.

Enter the Spaniard.

In 1887, Cajal traveled to a scientific conference and on the way he decided to pay a visit to Dr. Luis Simarro, a neurologist/psychiatrist in Madrid. (The distinction between neurologists and psychiatrists was unclear in those days and rightfully so. After all, every psychiatric condition *is* a neurological condition, but I digress.) Dr. Simarro was also an anatomy/histology enthusiast; he experimented with Golgi's technique and was able to improve it. He showed Cajal a series of slides using his modified Golgi method. Simply stated, Cajal fell in love with the technique, and predictably, he went to work enthusiastically (obsessively even) to further develop and improve the black reaction staining technique. Within a year, he published the first of many papers applying his own modifications of the Golgi method, significantly expanding the knowledge of the fine structure of the nervous system. Just to give you an example of Cajal's scientific productivity, between 1887 and 1891, he published about forty-five papers on the topic, an average of about eleven papers per year. Even by today's standards, this was a more than respectable number of papers, especially since Cajal did not have a large number of postdoctoral and graduate students working under him and did not have sophisticated imaging equipment or even a word processor!

Cajal not only contributed to the elucidation of the structural features of the nervous system but also significantly contributed to the understanding of how it works—in other words, its physiology. He formulated the theory of *dynamic polarization*, which essentially stated that the direction of transmission in a nerve cell is "one way only." Briefly, as we saw before, a signal is received by a neuron through its dendrites, then goes to its cell body, then to the axon, then to the nerve terminal, and finally to the next neuron's dendrites. This implied a *directionality* of nerve transmission. Cajal's physiological insights were remarkable, especially since he came up with them based exclusively on purely structural, rather than physiological, observations. Over time, the work of Cajal and others to establish that the neuron is the basic unit of the nervous system was articulated as the *neuron doctrine* by Waldeyer-Hartz and others.

A particularly ironic twist related to this story is that Golgi's black reaction was not widely used in histology until Cajal came along. In fact, Cajal probably was the scientist who did the most to improve and popularize Golgi's technique. This is indeed amusingly ironic since Golgi never publicly accepted the neuronist theory and remained the reticularists' very vocal champion, even when most of the scientific community adhered to the neuronist interpretation. The most publicized episode of the enmity between the two schools of thought took place in 1906 at, of all places, the Nobel lectures.

Golgi and Cajal were the 1906 awardees of the Nobel Prize in Physiology or Medicine. Historically, 1906 was the first year when the physiology or medicine prize was awarded to more than one researcher. Traditionally, Nobel Laureates deliver a lecture based on the research that got them the award. This is a high-profile event, usually attended by all kinds of dignitaries including scientists, politicians, and even the royal family of Sweden.

Well, there is no other way of saying it: Golgi essentially used his Nobel lecture to attack the neuron theory in a rather, let's say, *undiplomatic* way. The very first sentence of the lecture transcript reads:

> It may seem strange that, since I have always been opposed to the neuron theory—although acknowledging that its starting point is to be found in my own work—I have chosen this question of the neuron as the subject of my lecture, and that it comes at a time when this doctrine is generally recognized to be going out of favour.[8]

And things went really downhill from there, as Golgi essentially spent his lecture time attacking and trying to refute the neuronists' views, again, in a rather indelicate way. In all fairness, he was right when he stated that the development of the neuron theory was based on his work; nobody questioned that, as he did discover the black reaction after all. Moreover, everyone acknowledged Golgi's contributions to the field, first and foremost Cajal himself. On the other hand, he could not have been more wrong when stating that the neuron theory was going out of favor—anything but! Ironically, at the time, Golgi was for all practical purposes alone in his defense of the reticular theory.

However, Golgi's worst "scientific sin" was that in his lecture he did not properly refer or give due credit to his scientific forerunners and colleagues, even when many of them also contributed to the discipline. Some of them were even part of the audience at the Nobel lecture! To say that this was an uncomfortable situation is an understatement. In contrast, here is a transcript of Cajal's first sentence of his lecture:

> In accordance with the tradition followed by the illustrious orators honoured before me with the Nobel Prize, I am going to talk to you about the principal results of my scientific work in the realm of the histology and physiology of the nervous system.[9]

And then Cajal proceeded to deliver an actual scientific presentation, graciously paying due credit to his scientific predecessors, first and foremost Golgi himself.

This is how it is done.

To give this story some perspective, it is important to understand that Golgi ceased working on the nervous system in 1883 and completely stopped doing scientific research at all around 1889, dedicating his time mainly to politics, while Cajal continued working and publishing until a year before his death, in 1934. In this sense, Cajal represented the active scientist while Golgi represented the established, yet no longer active scientist. In many cases, not only in science, this inactivity results in the "comfort of familiarity," which makes it more difficult to accept newer points of view. This has nothing to do with age, as Cajal and Golgi's age difference was only about nine years; rather, it has to do with an attitude toward new things. This, of course, did not excuse Golgi's lack of recognition of scientific colleagues.

Now, I like saying that biology is a science of exceptions, as very few things in biology are absolute. The actual nature of the structure of the nervous system is an example of that. There are several types of neuron–neuron connections, as well as connections between neurons and other kinds of tissue, that actually associate the cells in a much closer way than what the cell or neuron theories suggest. For example, there are structures between cells called *gap junctions*, which are formed by certain proteins that allow for very fast cell-to-cell communication. These junctions allow the equilibration between two distinct cell environments. Also, there are certain places of the nervous system where the boundaries between neurons are not as well defined, just as Golgi asserted. Another peculiarity of the nervous system is that in some cases, the direction of transmission between neurons is not exactly "one way only." There are, for example, axon–axon connections (think about it in terms of a lateral street or shortcut). Also, in many cases neurons release neurotransmitters that do not necessarily act upon neighboring neurons, which is reminiscent of hormonal physiology. In fact, sometimes these neurotransmitters even act on the same neuron of origin, as part of feedback mechanisms. All these facts provide additional layers of complexity to the neuron theory. This means that even though the neuron theory still stands as the best explanation by far, in some cases, reticularist ideas also apply. Thus, Golgi was at least partially right after all. There was no need to disrespect anyone.

It is worth mentioning that both Cajal and Golgi were not quite full-time scientists, at least at the beginning of their careers. They were both physicians whose main occupation for a significant part of their lives was the clinic and the teaching and training of medical students. This makes their achievements all the more remarkable.

We have explored some fundamental aspects of the cellular structure of nervous systems; now, what about how they work?

EXCITABLE CELLS AND ELECTROPHYSIOLOGY

Neurons are capable of communicating with other neurons and other kinds of cells because they belong to a class of cells called *excitable cells*. Other examples of excitable cells are muscle cells, like the ones present in the heart, blood vessels, and digestive system, as well as all other muscles in your body. The term *excitable* here means that they can alter their internal chemical composition in a highly controlled way and use that change to communicate with neighboring cells. How do they do this?

All cells, excitable or not, must keep their internal and external environment isolated from each other. This is essential to sustain life. One aspect of this isolation involves the difference between the type and concentration of ions in or out of the cell (ions are electrically charged atoms, positive or negative). In neuronal cells, the two major players are sodium and potassium ions, with calcium as a prominent supporting actor. Under normal circumstances, there is an unequal concentration of such ions in and out of the cell. Since ions carry an electrical charge, these unequal environments render either side more positive or negative with respect to the other, depending on the specific circumstances.

As we mentioned in chapter 2, all cells are defined by the presence of lipid-based membranes, which separate the cell's external and internal environments. Since membranes are mostly lipids, they present a very efficient barrier for many types of substances, including charged molecules, like ions. In general, substances that easily dissolve in water-based media are called *hydrophilic* (water loving). In contrast, when a substance does not easily dissolve in water, it is termed *hydrophobic* (water fearing). We can easily visualize the difference between hydrophilic and hydrophobic substances by using the example of a simple vinegar/olive oil salad dressing. If you shake the salad dressing bottle, pour some on your salad, and set the bottle on the table, after a short while you will see that the liquid has separated into two distinct layers (called phases). The upper and therefore lighter layer corresponds to the hydrophobic part (oil), and the lower part corresponds to the hydrophilic part (vinegar, which readily mixes with water). The lipid-based membranes found in cells have the capacity of controlling which substances can go through it.

In general, charged substances like ions cannot cross membranes, as ions tend to be quite hydrophilic. Other, less hydrophilic substances have an easier time crossing the membrane. Since a cell membrane is able to select the substances that it lets go through it, we say that membranes are *semipermeable* or *selectively permeable*.

Since ions are generally not free to cross biological membranes, what causes some ions to accumulate in one side of a cell's membrane versus the other side? Many types of proteins are present within membranes; some of those proteins serve as gates that control the passage of particles across the membrane. One of these proteins is a very interesting macromolecule present in virtually every cell, called either the sodium-potassium pump or the sodium-potassium ATPase (from here on, we will call it the "ATPase" for short).

The ATPase is able to capture and break down adenosine triphosphate (ATP), the ubiquitous energy currency of life (see chapter 2). When the ATPase breaks down ATP, some of the energy liberated by this breakdown is channeled to power up physiological events, somewhat similarly to when gasoline is ignited and burned under controlled conditions in a car's engine, where some of that energy is channeled and transformed into mechanical energy (movement).

In the case of the ATPase, this protein uses the chemical energy in ATP to power up the transport of three sodium ions out of the cell, more or less simultaneously with the transport of two potassium ions into the cell. It is important to point out that both sodium and potassium are positively charged ions, so essentially we can say that the ATPase modulates the transport of three positive charges out of the cell per two positive charges that get in. This allows us to compare this process with the flow of money in a bank account. For example, if you deposit two dollars (two potassium ions) for each three dollars you withdraw (three sodium ions) in your account, pretty soon you will owe money to the bank; you will be in the proverbial red, or in the *negative* if you will. Thus, by convention, in electrical terms we can consider the inside of the cell more *negative* than the outside. Please note that this is a simplified representation; many ions besides sodium and potassium are involved in this process, including magnesium, chloride, and so forth.

What does this mean in a physiological sense? When there are different ion concentrations in and out of the cell in the way described previously, we say that the cell is *polarized*, essentially meaning that whatever is "in" is not identical with whatever is "out," specifically in terms of electrical charge. All cells display this polarization property, not only excitable cells. In the case of excitable cells, we call this polarization the *resting potential*, which can be visualized as a voltage difference between the cell's insides and its external environment. Again, remember that we are using sodium and potassium in this description for simplicity's sake; even though sodium and potassium are the main players, many other ions play a role in this process as well. The actual magnitude of the resting potential can be estimated by taking into account all the other ions.

Please know that I am fully aware that all this talk of ions and so forth may sound a little dry and obscure at times, but these are the mechanisms that actually help you function as a living organism. Without them, you would not be alive. Yes, they are that important. Further, they will help you better understand many of the main points of the book. Hang in there, you are doing great!

About the voltage difference between the inside and outside of neurons, it is important to realize that these are really small voltages, measured in the millivolt (mV) range (1 mV is 1/1000 volts). For comparison, consider that normal batteries, like the ones in a remote control, usually carry a charge of about 1,500 mV (1.5 volts) and the common electrical outlet in the United States is about 110,000 mV (110 volts). It is truly remarkable that excitable cells use these tiny voltages within them to transmit information.

Excitable cells are capable of a really nifty trick to use changes in the resting potential to communicate with other cells. They use specific proteins in their membranes, called *voltage-gated ion channels*, that, when opened, allow for ions to flow across the membrane according to their *concentration gradient* (this is just a scientific way of expressing how many particles are present in a given volume in each side of the membrane). This ion flow changes the magnitude of the resting potential. When the voltage changes toward a more positive value, we call this phenomenon *depolarization* (less polarized), which can be seen as an electric current. In contrast, when a resting potential equals exactly zero, there is no polarization at all. In other words, the type and number of electrical charges in or out of the cell are equivalent.

Many living organisms use this change in the cell's electrical potential to transmit information from one cell to another. When an excitable cell transmits information in this way, this process is called the *action potential,* which is the basis for the discipline of electrophysiology.

SYNAPSES AND CHEMICAL NEUROTRANSMISSION

But what causes the action potential and what does it do? The points of contact of neurons with other neurons or with muscle cells are specialized structures called *synapses* (Figure 3.3). The term *synapse* was coined by another neuroscience pioneer, Charles S. Sherrington. Interestingly (and fortunately), he came up with the term at the suggestion of Arthur W. Verrall, a humanities professor. This is fortunate because Sherrington's original word choice was *syndesm.*

Ouch.

Synapses come in two main classes: electrical and chemical. Electrical synapses are essentially the same as the gap junctions that we mentioned before. On the other hand, the chemical synapses are best understood in terms of physiology and pharmacology. Chemical synapses are the target of many medications, from muscle relaxants to antidepressants.

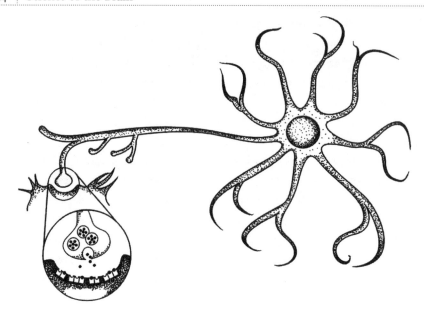

FIGURE 3.3. Typical neuron showing a chemical synapse. Illustration by A. G. Pagán.

Here's how a chemical synapse works. A typical synapse is composed of three main parts, a presynaptic side (part of one neuron), a postsynaptic side (part of a second neuron), and a space that separates them, called the *synaptic cleft* or the *synaptic gap* (Figure 3.4). This is a tiny space; the length of a synaptic cleft is close to 0.05 microns (μm). A micron is a millionth of a meter.

One of the consequences of the action potential in animals is the release of chemical substances from the presynaptic neuron; these molecules are stored within membrane-bound structures called *synaptic vesicles* (Figure 3.4). When these synaptic vesicles get close to the presynaptic membrane side, they fuse with it and release substances that diffuse across the synaptic cleft and eventually bind to postsynaptic receptors. These compounds are called *neurotransmitters* (Figure 3.5) and they control virtually every physiological aspect of the organism in one way or another. There are more than a hundred types of neurotransmitter molecules identified. Even though the most abundant one is glutamic acid (glutamate in its active form), the best-known one is *acetylcholine,* which was the first neurotransmitter discovered and has important roles in the muscular and nervous systems. Other neurotransmitters include *serotonin*, which has a well-established role in conditions such as depression, and *dopamine,* which when deficient is related to pathological conditions such as Parkinson's disease and is also related to addiction and drug abuse behaviors (addiction and drug abuse are

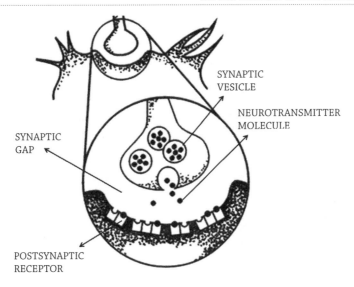

FIGURE 3.4. A chemical synapse.
Illustration by A. G. Pagán.

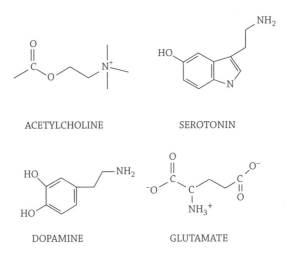

FIGURE 3.5. Several neurotransmitter examples.
Illustration by O. R. Pagán.

not necessarily the same thing; you can get addicted to things other than chemicals—think gambling or even sex, for example).

At the receiving end, there are proteins to which these and many other neurotransmitters bind; these proteins are called *receptors*. These receptors perform a wide variety of roles within cells, mostly related to signaling. Some of these receptors regulate cellular communication events through second messengers; these are generally called *metabotropic* receptors. Certain kinds of receptors regulate cell-to-cell communication through ion channels; these are called *ionotropic* receptors.

Ion channels are one of the most important classes of molecules within nervous systems. One of their main characteristics is that most ion channels are selective; they allow some ions but not others to go through membranes. In normal neuronal physiology, when a sodium-selective ion channel opens in a neuron at rest, sodium flows in, since there is more sodium outside than inside. This will slightly change the resting potential, making it less negative. That change in voltage in turn induces the opening of more voltage-gated sodium channels, which changes the voltage even more. At some point, there will be a sudden, rather than gradual, voltage change, spiking it to positive voltages, staying positive for about a millisecond and then going back to its initial level (this is an action potential in a nutshell). In turn, the action potential triggers a wave of calcium ions into the cell, which then modulates the release of neurotransmitter substances in the synaptic cleft. The binding of neurotransmitters to their receptor targets can trigger the next wave of action potentials. In this case we talk about *excitatory neurotransmission*. In other cases, the binding of the transmitter to the receptor makes the neuron less active; in this case we speak of *inhibitory neurotransmission* because action potentials are less likely to happen. This is, of course, a very simplified description of the synaptic physiology process; there are many variations of the theme depending on the specific cell system, but I hope that you get the general idea. Again, this is a simplified view of a rather complex process, a process so important that it allows for all the functions of nervous systems to work properly.

Any well-coordinated process needs to be terminated at some point; if not, it would be like having a car with a gas pedal but no brakes. There are five described ways of terminating neurotransmission.

The first one is simple diffusion, where the transmitter molecules drift away in the extracellular media. A second way is through the breakdown of neurotransmitters by specialized proteins. An additional way is through "scavenger" proteins floating around the synaptic cleft; these proteins bind to the transmitter molecules, effectively taking them out of the way. Most receptors also undergo a process called *desensitization*. This means that over a certain period of time, the receptor becomes insensitive to the transmitter's presence, in essence terminating neurotransmission. The last way to terminate this process is through proteins located at the presynaptic site, called neurotransmitter transporters. These proteins modulate the uptake of neurotransmitter molecules to the presynaptic side, where they are either destroyed or repackaged into vesicles for the next transmission event.

There are many types of drugs, legal and otherwise, that affect synaptic physiology, often at the termination mechanisms just outlined. Many of these drugs mimic the structure of neurotransmitter molecules. In some cases, they enhance neurotransmission, and in other cases, they inhibit it. Also, due to the physiological importance of such systems, many organisms produce toxins that target one or more neurotransmission steps. Please recall the conotoxins that we saw in the previous chapter. Another example is botulinum toxin, one of the nastiest toxins known, which interferes with the release of acetylcholine from the synaptic vesicle in a nerve–muscle synapse, therefore inhibiting transmission. You may have heard of botulinum toxin as *Botox*, mainly used in cosmetic procedures. Other compounds affect neurotransmission termination; these include medications against depression, like the famous Prozac, and abused drugs like cocaine and amphetamines, among others. These compounds work by interfering with neurotransmitter reuptake, therefore enhancing transmission.

IS A NERVOUS SYSTEM ABSOLUTELY NECESSARY FOR SURVIVAL?

In this chapter, we have explored some aspects of the importance of nervous systems for activities such as learning, which improve the survival chances of living organisms. However, an important question would be whether nervous systems are actually necessary for survival. This is, of course, a rhetorical question. Many single-celled organisms thrive and are actually quite successful without a nervous system. Moreover, multicellular organisms with a high degree of complexity (plants and fungi) also lack nervous systems and are doing very well, thank you very much. It is important to take into account that if we count plants, fungi, and microorganisms, we are talking about the vast majority of life forms on earth. Therefore, we "nerve bound" organisms are a true minority!

Despite being a mere fraction of this planet's wonderful biodiversity, we cannot deny that as far as we know, only organisms with a nervous system of a certain level of complexity possess the potential to express a quality that is very close to our hearts: curiosity. That curiosity is one of the main traits that allowed us humans to survive as a species and that eventually provided the impetus for us to try to understand our universe. In a very real sense, this same curiosity may hold the key to our collective survival. As far as we can tell, only humans wonder; only we seem to think about the universe; further, we often wonder about our place in it. This is one of the reasons the study of the nervous system is so fascinating.

The development of nervous systems did not happen in a vacuum; our evolutionary history shows that many of the components in terms of structure and function that are essential for nervous systems are already present in other life forms that do not possess one. For example, many unicellular organisms like paramecium use neurotransmitterlike molecules and even molecules that are actually used as neurotransmitters in "higher" animals as signaling molecules. These molecules in turn allow for behaviors that make paramecia and other unicellular entities swim toward food or swim away from toxic environments. Further, these behaviors of microorganisms are controlled by electrophysiological phenomena similar to the ones we mentioned before, like the action potential. In many cases, these properties provide clues about the evolutionary history of these physiologies.

Incidentally, this is yet another example that invalidates a claim that is frequently raised against the evolutionary view of life. Simply stated, the argument is "What good is half an eye?" which means that to be of any use at all, a complex physiological system like the eye, for example, must possess all its component parts. In other words, it either works or does not work, with no middle point. This could not be more wrong. Many single-celled organisms possess patches that respond to light; these are little more than an accumulation of photosensitive pigments, lacking lenses or any of the structures that we associate with an eye. Granted, they are not able to form any kind of image to speak of, but the slight edge provided by the ability to distinguish between light and dark can be the difference between survival and death.

The same argument applies to the individual cellular and molecular components in simpler organisms that eventually form part of nervous systems in more "advanced" ones.[10] An amoeba, for example, does not have any nervous system to speak of, but a series of signaling molecules help them to survive nonetheless. Further, these signaling processes have been found to be some of the main players in cell-to-cell cooperation, as illustrated by slime molds, which are essentially social amoebas, considered by many as an intermediate step between unicellular organisms and multicellular animals.[11]

There are examples of these biological communication phenomena in more complex, multicellular organisms. Take plants, for example. If we look at it from a certain perspective, we can argue that plants in a certain sense invented neurobiology! Of course, it is not that simple; let's talk about it. Behavior is usually associated with responses to environmental stimuli. These responses are usually associated with changes in the organism; an evident sign of these changes is some kind of motion. In general, motility is usually associated with animals; however, make no mistake, plants display behavior, and oftentimes this behavior is expressed as motility as well! Think about the sensitive plant, *Mimosa* sp., which is capable of rapidly closing its leaves when touched, and even more dramatically, think about the Venus flytrap, *Dionaea muscipula*, which actively captures insects and even the occasional small vertebrate. These and other *carnivorous plants* are actual predators! People have always been fascinated by carnivorous plants, precisely because of their unusual behavior of capturing other organisms for nourishment. Again, keep in mind that these behaviors in carnivorous plants can be considered to some extent hunting behavior, which is usually associated with animals. The first formal study on insectivorous plans was published by (who else?) Charles Darwin, in 1875. He noticed close parallels between carnivorous plants and animals; these similarities piqued his curiosity to no end. He was particularly taken by all the parallels he observed between these plants and animals. Even Darwin's wife, Emma, ever the supportive spouse, whimsically commented on her husband's fascination with the Venus flytrap, "I suppose he hopes to end in proving it to be an animal."[12]

It is pretty evident that plants can express behavior; however, plants have no nervous system that we can recognize. How can they generate behavior?

PLANT NEUROBIOLOGY

Yes, this is an actual field of study, albeit one with a tortuous conception, particularly painful birth, and a very difficult childhood as we will see. It has not reached its adolescence yet, and as we all know, puberty is an especially long and painful process, particularly for parents (bad joke).

Plants have no neurons or, for that matter, no structural features recognizable as nervous systems (at least macroscopically), but, ironically, the etymology of the very word *neuron* roughly translates as "vegetal fibers." The names of some other neuronal parts also share their roots with terms associated with plants. Take dendrites, for example; the word implies a treelike structure (as in *dendrology*, the study of trees). Despite having no neurons, many types of plants possess cells that contain subcellular machinery that seem to work in a similar way to neuronal cells. In fact, many plants produce substances that are bona fide neurotransmitters in animals, such as acetylcholine, serotonin, and glutamate, among others (Figure 3.5). Moreover, plant cells can generate certain types of activities suspiciously similar to action potentials! This was discovered in 1873 by an animal physiologist, John Burdon-Sanderson, who worked on the Venus flytrap instigated by Charles Darwin.

It is not entirely clear why or how, but the interesting and promising field of plant electrophysiology lost its appeal over time. Some authors think that this was due to an increasing interest in chemical means of communication. At any rate, the 1970s brought more evidence pointing to electrical signaling as an important phenomenon in plant physiology. However, in an almost fatal blow, the growing interest in these kinds of phenomena was slowed down and eventually reversed by the publication of the book *The Secret Life of Plants* by P. Tompkins and C. Bird.[13] This book was not generally accepted as actual science and was heavily criticized by the scientific community.[14] The eventual consequence was that this book substantially diminished the enthusiasm for the scientific study of plant electrophysiology, and many scientists understandably distanced themselves from a field that some associated with pseudoscience. This was rather unfortunate; it is indisputable that plants display electrophysiological properties and that these properties play an important role in the adaptation and survival of plants in challenging environments.

Then again, this resilient field refuses to die. Its latest incarnation (although still highly controversial) is now called "plant neurobiology," which is currently intensively studied. As we saw, the terms *neurobiology* and *behavior* are historically associated with animal life, whose behavior is generated by nervous systems. At their most basic level, nervous systems must be able to change their fundamental properties in response to environmental stimuli. Please recall that the main mechanism used by nervous systems to achieve this is through the generation of action potentials.

Controversy is not uncommon in scientific fields, especially in newly developing or even reimagined areas. Plant neurobiology is no exception. A lot of good research is being done, but many published papers in this area have been criticized as overreaching with their conclusions when interpreting experimental results. Additionally, there are some challenges to the interpretation of such results; further, some of these results are dismissed as experimental artifacts by highly qualified experts in the field of electrophysiology. At least a fraction of the lack of credibility of this area of research is due to the fact that some researchers in this field chose the term *plant neurobiology* and phrases like *brainlike*, which undoubtedly has rubbed some traditional neurobiologists the wrong way. Some scientists dispute the use of these terms even as metaphors; part of their argument is that they do not give the impression of seriousness, precisely because plants have no neurons! Additionally, some plant neurobiologists are not helping their cause by starting to talk about plant "intelligence" and "cognition." The terms *intelligence* and *cognition* are not fully understood and are very controversial concepts in their own right, even when viewed exclusively from their traditional perspective of animal biology.

Science is a process performed by people, with all the biases and imperfections inherent to human behavior. It is entirely possible that some scientists dismiss the claims of the plant neurobiology field at a visceral level, just because of a psychological barrier due to the way the field is named. This controversy has generated opinion (as well as opinionated) articles with colorful titles like "Plant Neurobiology, Intelligent Plants or Stupid Studies,"[15] which, understandably, is not taken kindly by advocates of the field.

Despite all these controversies, I believe that the study of plant electrophysiology, plant neurobiology, or however they want the area to be named is a legitimate, very interesting research field, with the potential of producing significant scientific results. A very healthy sign is that this proverbial battle is being fought primarily via scientific journals, as it should be. The tools of classical neurobiology, in conjunction with current advances in molecular biology, should make the next few years a very interesting period for plant behavior research. To learn more about the current state of aspects of plant behavior research, the recent book *What a Plant Knows* by Dr. Daniel Chamovitz provides an excellent introduction on how plants perceive their environment and how they react to it.[16]

As we've seen, when many neurons organize themselves in a way that allows them to communicate with each other in a coherent manner, as in the events described in this chapter, very interesting things happen, like behavior, for example. Currently, the most complex structure of this kind that we know of is the human brain, which we'll talk about next.

Shaped a little like a loaf of French country bread, our brain is a crowded chemistry lab, bustling with nonstop neural conversations.
—DIANE ACKERMAN

4

THE HUMAN BRAIN

WHAT EXACTLY IS A BRAIN?

TO ANSWER THIS question, we must first clarify what kinds of organisms we are talking about and we further have to decide whether to define the "brain" in terms of structure, function, developmental biology, a combination of all three, or an altogether different criterion or criteria. Not unexpectedly, our first tendency would be to take the vertebrate, specifically human, point of view as a starting point. That said, now more than ever, it is widely recognized that at least in part the rudiments of thought and even consciousness are present in "lower animals."[1] In many of these animals, however, the structural and, to some limited extent, the physiological definitions of a vertebrate-type brain will not apply.

With respect to the vertebrate brain, there are basically five "theories of brain architecture"[2,3] that admittedly in an arbitrary way study the vertebrate brain in terms of:

- Aristotle's concept of a "dual brain" composed of the encephalon (the brain) and parencephalon (roughly, the cerebellum)
- Segmental organization, which includes the ten basic parts of vertebrate brains that are currently recognized: cerebral cortex, basal ganglia, thalamus, hypothalamus, tectum, tegmentum, pons, medulla, cerebellum, and spinal cord
- Developmental origin

- Evolutionary history
- Genomics

Keep in mind that as with everything biology, these schemes are logical yet arbitrary and have the potential of being modified. Also, please note that these classifications are mainly about structural features as opposed to physiological features. As we saw, the study of complex systems through the study of their component parts is an established scientific practice. Finally, in practice, these points of view do not need to be mutually exclusive. We can study the brain using more than one of these theories in any combination, or we can even use all of them for that matter. In fact, all points of view combined will paint a more reliable picture of what we want to understand. Remember that nature does not care about our classification schemes; nature just is.

Again, it is important to keep in mind that none of the aforementioned schemes deal directly with function. This is an important point that we will explore later, but in the meantime think about the following: is a brain best described by what it physically is or by what it does?

After the cellular level, vertebrate nervous systems are composed of clusters of neurons. The cells in these clusters communicate with other cells and with other clusters to perform specific functions. When a neuron cluster like this is located outside of a central nervous system, we call it a *ganglion* (plural: *ganglia*). On the other hand, such clusters are called *nuclei* when located within the central nervous system (singular: *nucleus*; different from a cell's nucleus). This is a formal distinction that anatomists agreed upon only after some parts of the human brain were already named and described, which led to some confusion. For example, there are parts of the brain like the *basal ganglia* that are actually located within the central nervous system and therefore in strict terms they should be called *basal nuclei*, but any neurobiologist knows what they are either way.

The question is, is a brain, then, a collection of interacting nuclei? If we assume that this is true, then the next logical question is: what is the minimum number of nuclei required to make a brain? In other words, when does a ganglion become a brain?[4] If we expand this discussion to include nonvertebrate brains, it will become evident that a brain can take various forms, from the bilobar brains of flatworms and humans to the magnificent yet "weird" brain of the octopus, which wraps itself around the animal's esophagus.

What about function? Can we simply say that "a brain is as a brain does"? Functionally, a good working definition of a brain is as "the control center of a biological system." But let's not forget that there are many organisms that have a nervous system, albeit not a centralized one, and they not only survive but also even thrive.

An acutely insightful and quite witty exploration of the multiple meanings that the concept of a "brain" can have was published in 1986 by the late Dr. Martin G. Netsky,[5] and I believe that his thesis is still valid today. Essentially, the article posits that an all-encompassing definition of what a brain may be is a very difficult thing to articulate, if at all possible. This is so because there are multiple ways to see a brain such as the ones I alluded to before, namely, what it is, what its component parts are, and what it does. An early definition of the word *brain* (by Samuel Johnson in 1755 as noted in Dr. Netsky's article) was "that collection of vessels and organs in the head from which sense and motion arise," which is as good a definition as any, even now.

THE HUMAN BRAIN AND NERVOUS SYSTEM

The human nervous system, as with every other biological system, is the result of evolutionary forces. We will talk in more detail soon about the evolution of nervous systems in general. In the meantime, let's talk about structure and function.

Anatomically, the nervous system in vertebrates is subdivided into two main parts, the central nervous system (CNS for short), which includes the brain and spinal cord, and the peripheral nervous system (PNS), which includes essentially all other nervous tissue not included in the CNS (Figure 4.1).

The simplest starting point for describing a human brain is to talk about its size and mass. The "three pounds" phrase, albeit frequently used, is ideal to provide the initial perspective that truly sparks our sense of awe for the organ. How can a mere three pounds of organized tissue do everything we know the human brain does, let alone all the many things that we surely do not know it does?

When looking at an intact human brain, the first thing that one notices is that it seems to be a double organ (Figure 4.2, left). That the brain is a double organ in itself is not remarkable; after all, many of our organs come in pairs, like the lungs, the kidneys, the eyes, and so forth. In those cases, though, each organ, left or right, performs essentially the same function as the other one. In contrast, the two brain halves (or hemispheres), although pretty symmetric, do seem to have a slightly different structure and therefore function. However, keep in mind that the human brain is surprising plastic. This simply means that the structural features of the brain are not static in the sense that they can be reorganized at least to some extent. For example, in many cases when a part of the brain is damaged by accident or disease, undamaged parts of the brain can reorganize and take over some of the lost functions.

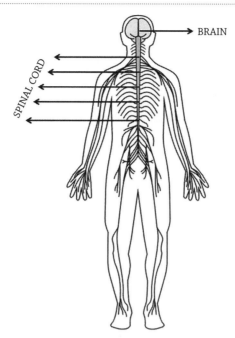

FIGURE 4.1. The human central and peripheral nervous system. The peripheral nervous system was simplified for clarity, as nerves reach every region of the body.
Illustration by A. G. Pagán.

At any rate, the brain halves that are the structure that we think of when we think of the word *brain* are only part of it; the hemispheres are formally called the *cerebrum*, which would be the visible part of the so-called cerebral cortex, which seems to be a central part of our brains. Strictly speaking, the human brain is formed by the cerebrum, the *cerebellum,* and the *brainstem* (Figure 4.2, center). In a very simplified way, we can say that the cerebrum deals with thought, the cerebellum deals with the control of movement, and the brainstem deals with complex automatic functions like cardiorespiratory physiology and the sleep–wake cycle, among other tasks. As I said, this is a very simplified and general view; for example, the cerebellum, in addition to its role in motor control, is also involved in cognitive processes.

The next thing we notice when looking at the cerebrum is that it has a corrugated appearance. This is also true of the cerebellum. In ancient times these corrugations were compared to melted metal, and that is not a bad comparison at all. However, these rumples are anything but random; they have a distinct structure that is pretty much conserved from one brain to another in the absence of pathologies or developmental abnormalities. In other words, these structures are very similar, but not quite identical, in all brains. These grooves are divided into wrinkles or folds, also known as *gyri* (singular: *gyrus*); the boundaries between

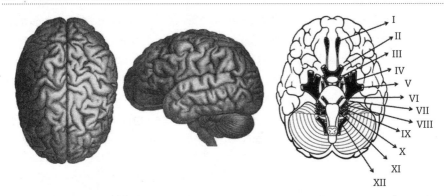

FIGURE 4.2. *Left, middle*: The human brain, top and side view. (From The Soul of Man, by Paul Carus, Open Court Publishing, 1905. Public domain; downloaded from Wikimedia Commons.) *Far right*: The human brain viewed from below, showing the cranial nerves.
Illustration by A. G. Pagán.

gyri are known as *sulci* (singular: *sulcus*). The main purpose of this arrangement is ostensibly to increase the overall surface area available to the cerebral and cerebellar cortexes, so they can be accommodated in a relatively small volume. It is a little bit like these normal size T-shirts that are packed in a one-cubic-inch volume. The tight packing allows for the use of a much smaller volume. These sulci and gyri also serve as anatomical landmarks that help us localize regions with defined functions.

As expected of a brain, the human brain is the control center of the body. To that end, there is a series of nerves that control various sensory and motor functions. We call them the cranial nerves. These nerves relay sensorial or motor signals to various body parts (Figure 4.2, right).

Based on what we have seen so far, we can briefly describe our brain in terms of structure and function. Basically, the human brain is a bilobar structure, as represented by the two hemispheres, which can be subdivided into distinct regions that control specific functions. Some of these functions are channeled by the cranial nerves, which provide the link between structure and function; in turn, these functions include motor and sensory processes. The understanding of this basic arrangement is the starting point for research aimed at the development of treatments and medications to treat neurological conditions. We now understand the central role of our brain, but historically, that was not always the case. In the next section we will briefly talk about some aspects of how the brain was perceived by people over history.

A BRIEF HISTORY OF WHAT PEOPLE THOUGHT OF THEIR BRAINS

The human brain is frequently hailed as the most complex structure that we know of, and rightfully so. Our brain is the point of origin of everything we do, all the things we associate with the human condition. To the best of our current knowledge, love, hate, everything in between, and even consciousness itself seem to arise from the highly coordinated activity of roughly 100 billion (or so) nerve cells within our heads. Understandably, we really keep the brain in high regard; one of our reasons for doing so is that, in a very real sense, our brain is us.

However, in ancient times the importance of the brain was not immediately appreciated. In practically all early cultures, not the brain but the *heart* was considered the true control center of the body. Not only that, the heart was also thought of as the seat of emotions and the dwelling place of the soul itself. I don't know about you, but sometimes our ancestors' point of view puzzles me somewhat; allow me to explain.

On the one hand, I do understand why earlier humans related our internal organs with emotions. After all, this link is immediately apparent. We are all familiar with the very direct connection between our body and our feelings. Anyone who has ever felt fear knows that the heart beats faster and many, if not all, of us frequently experience the "butterflies in the stomach" effect when nervous. This is normal and even essential to our survival; you may recognize these reactions as part of the well-known fight-or-flight response. We even have "gut feelings"—yes, they are more real than we realize! Our digestive system has a rather complex nerve network, which has been likened to a "little brain" or a "second brain." This is a very intensively researched subject nowadays, since it is clear that our "second brain" can affect our behavior, too. So, in light of these facts, it is understandable that our forbearers associated their mental activity or at the very least their emotions with their "guts," so to speak. Fair enough.

On the other hand, when I think (which is most of the time), I feel my thoughts *inside my head,* specifically, right behind my eyes. Am I weird? Why yes, yes I am, and so are you. I do not believe that anyone is 100% normal (in strict terms, nobody really is), but this is a topic that we will not discuss in this book.

Regarding your very own thought processes, I strongly suspect that you also feel your thoughts inside your head if you think about it (pun absolutely intended). Furthermore, that the brain was not considered a very important organ in early cultures is also perplexing because it is very clear, based on early written records, that ancient peoples knew the potentially devastating consequences of head injuries. They were well aware that head damage due to accidents or war injuries frequently had an effect on body parts that seemed undamaged. Some early thinkers presciently entertained the idea of the brain as the center of thought even though this view was not generally held by the learned people of the day.

We have direct evidence indicating that ancient humans were very aware of the link between head trauma and undesirable effects. The oldest medical text

FIGURE 4.3. The first known instance of the word *brain* in any written language, ancient Egyptian. Illustration by A. G. Pagán.

that we know of, the *Edwin Smith Papyrus*, describes many examples of head injuries and their consequences. This ancient Egyptian document, which is thought to be more than 3,500 years old, is the very first known example of a treatise in trauma medicine. This document describes many cases of accidental injuries, not only to the head, but also to other body parts. In a way, it can be considered the first text on industrial or military medicine, since much of the damage mentioned there is thought to have come from descriptions of construction accidents or battle wounds. Interestingly, in addition to describing these injuries, the text also classifies them as "treatable," "probably treatable," or "untreatable," anticipating modern triage protocols in emergency medicine.

The *Edwin Smith Papyrus* is also important from the perspective of our narrative, as this papyrus contains the first-known written example of a word that meant "brain" (Figure 4.3).

You'd think that since the ancient Egyptians had a word that meant "brain," they had a certain measure of respect for it, right? Well, not quite. Literally translated, the word itself roughly meant "skull-offal," which sounds to me like "head tripe." To add insult to injury, please consider this: the process of mummification is one of the practices the early Egyptian culture is best known for; even though mummification is not found exclusively in the ancient Egyptian culture, we usually associate it with them. The ancient Egyptians' beliefs in an afterlife heavily influenced the development of mummification. Wealthy early Egyptians went to great pains to leave instructions and resources to, upon death, preserve many of the internal organs for their use in the next world. The not so wealthy could not afford to have their organs independently preserved but still tried to arrange to preserve their bodies with aromatic oils and certain salts. In any case, the brain was not preserved at all; the brain was actually scooped out of the skull through the nose using a special tool and then discarded—again, no respect whatsoever.

Things did get a little bit better, but not much, with the Greeks. Alcmaeon, who lived around 500–450 BC, first proposed the brain as the origin of mental function and perception. Plato (429–347 BC), probably the best-known ancient philosopher (and in my view the very best philosopher or any time period), recognized the nervous system as the site for rational, but not emotional, thought. Later on, Aristotle (384–322 BC) still associated mental states not with the brain, but instead with the internal organs. In this respect, there was no unanimous consensus, as other less influential philosophers still related the brain with thought. It was a step in the right direction, but still, the Aristotelian point of view was predominant. That said, in contrast with the early Egyptians, who did not seem to assign any kind of function to the brain, Aristotle at least thought the brain had a function. Care to take a guess as to what that function was? Aristotle thought the brain was a cooling system, essentially, a radiator.

To be fair, one can understand his reasoning, but only up to a certain point. In contrast with the heart and other muscles, the brain did not seem to move, twitch or pump anything. Aristotle reported that brains looked moist (true enough) and (strangely) that they felt cold to the touch (huh?). This is somewhat puzzling. We know that an important aspect of Aristotle's approach to nature is that he emphasized observation rather than pondering. The thing is, brains are rich in blood vessels; as a result, they have the same temperature as the body under normal circumstances. If Aristotle would have examined fresh brains, or even the brains of live animals, as the later philosopher and physician, Galen, did and reported, he would have noticed that animal brains, particularly mammalian or bird brains, were actually warm to the touch. Therefore, we now think that he must have only examined the brains of animals that were dead for some time (never human brains though, as there were taboos associated with the dead human body).

Even though Aristotle gave the heart the highest importance, the brain was a close second. He argued that the "hotness" of the heart was counterbalanced and controlled by the "coldness" of the brain.

It seems that another reason Aristotle did not consider the brain important was that many types of animals do not seem to have much of a brain, if at all, and they did not seem to care. These animals are generally small, like earthworms or starfish, for example; however, some jellyfish can get pretty big and Aristotle surely had the chance to study them. Virtually every type of animal of a certain size that Aristotle may have had access to (and it is well documented that he had access to a lot of them, all the way up to elephants)[6] possessed very self-evident brains or something like it.

Slowly, but steadily, things started to look up for the brain. After Aristotle's death, two physicians from Alexandria, Herophilus and Erasistratus, dedicated themselves to the study of human anatomy from the top of the head to the tip of the toe. These physicians were actually rumored to perform human vivisections (yes, this is exactly what it sounds like) using prisoners; however, this seems to be a made-up charge by jealous competitors. What is true for sure is that they actually examined human cadavers from recently executed prisoners. Herophilus and Erasistratus realized for the first time that nerves were entities distinct from, say, blood vessels. More importantly, they traced the nerves' origin to places like the base of the skull (as in the cranial nerves) and the spaces between vertebrae. However, they ascribed the origin of mental functions to the brain ventricles, which are cavities filled with cerebrospinal fluid.

With Hippocrates (460–? BC), the position of the brain as the control center of the body began to become mainstream knowledge. He wrote, "Men ought to know that from nothing else but the brain come joys, delights, laughter and sports, and sorrows, griefs, despondency and lamentations."[7] He also wrote on head injuries and recognized that a lesion on one side of the brain usually affects the opposite side of the body. Also, he considered epilepsy as an actual disease of the brain and not something inflicted by the gods. He also gave us many other insights on brain function.

From Hippocrates on, the brain was then pretty much established as the place of origin of the mind. The early history of neuroscience, specifically, the history of how the brain was explored over time, is far richer than the material covered in this brief overview. Many other philosophers, scholars, and physicians worked on filling the blanks in terms of the brain structure and, to a lesser extent, function, but we will not talk about them here. There are many excellent books that present a more complete exploration of the history of neuroscience should you wish to know more.[8-10]

Before I go on, I do not want to give the impression that I think less of all of these early explorers of the brain or that I am belittling their efforts—quite the contrary. Their works and achievements are truly remarkable, especially when we realize that they explored neuroscience and, in fact, all of nature as they knew it without the advantage of all the technology that is at our disposal now, from microscopes to computers. Only God knows what people living a thousand years from now will think of the science we practice today. That said, we have the distinct advantage of the experimental method, as discussed in chapter 1. Because of that, we are reasonably sure that we are on the right track research-wise. Additionally, science and technology are always developing, generating new ways of thinking and improved techniques and instrumentation. Neuroscience in particular is a very dynamic science that still has a lot to discover. Let's continue our exploration of nervous systems by getting an idea of how complex they really are.

THE COMPLEXITY OF THE HUMAN BRAIN/ON REALLY BIG NUMBERS

The human brain is composed of about 100 billion neurons. Interestingly, 100 billion is also a good estimation of the number of stars in a typical galaxy, and there are about 100 billion galaxies (give or take) in the known universe. Science is full of these wonderfully curious coincidences. We will now explore aspects of the nervous system from the perspective of the human brain as a biological organ. I want to give you an idea of how incredibly complex the brain actually is and how, despite this complexity, scientists can study it.

Lately, some popular authors speak of the brain's "86 billion neurons," based on research that examined exactly four brains where they determined an average of 86 ± 8 billion neurons.[11] This study, although very well done, needs to be replicated, if possible with a larger sample (meaning more brains). That said, whether there are 78, 86, 94, or 100 billion neurons in a typical human brain makes essentially no difference to the point I am trying to make.

Frequently, we speak of billions (and even trillions in these difficult economic times) without quite understanding what these numbers really mean. Let's take the 100 billion neurons in a typical human brain that we just alluded to as an example. Think about the following: what does it mean to have 100 billion of anything? Just so you know, in our discussion 100 billion equals 100,000,000,000.

Allow me to try something that may help us visualize a little better the real magnitude of these kinds of numbers. Let's start with 1 billion (1,000,000,000 in American English). Suppose that I ask you, how long will it take to count to a billion? (This is the part where you get cute and say "A billion seconds!" More than one student has pulled that one on me.)

True enough, it may take a little bit more than a billion seconds, but of course, this does not help us to understand the concept of a "billion." Our minds are simply not equipped to deal with such quantities in our daily lives; in other words, we cannot visualize the real immensity of these numbers.

For example, let's suppose that you are hungry, not just hungry for anything, but hungry for something really, truly delicious. You find the recipe for *paella* (a rice-based Spanish dish; yum!) and decide to try it. When preparing it, you do not count the grains of rice that you need as per the recipe's instructions. Rather, you use units of measurement that your mind can manage, like a pound, a kilogram, and so forth. In other words, if the recipe for paella (yum! If you have ever tried it, you know why I just *have* to say it again) requires a pound of rice, you would certainly not count about 32,000 grains of rice (yes, I did the calculation); you would measure about a pound of rice instead.

So let's try this again: how long will it take to count to a billion, say, at a rate of one number per second, nonstop?

Anyone? Anyone at all? Don't be shy....

It is actually a simple calculation, the kind of calculation that you can do (literally) on a napkin while having a cup of coffee. It would take about thirty-two years. Yes, that long. Also, please keep in mind that it would take that long counting twenty-four hours a day, seven days a week, nonstop. Now, please remember that a normal human brain has about 100 billion neurons, so to count each of these neurons, we would need about 3,200 years!

Hang on; it's going to get better....

Neurons must communicate with other nerve cells at specific points of contact (the synapses—see chapter 3); this is the name of the game when generating brain function. The sheer number of cells, as important as it is, is not enough; 100 billion identical, unconnected (or connected in a random way) neurons would not make a brain. An essential feature of a brain is that its component cells must also be able to communicate with other neurons. A typical neuron in a brain has about 1,000 synapses, but it can have up to 80,000 depending on the neuron type! The average seems to be 30,000, more or less according to most references, so let's take this average as our starting point.

Given this information, can you imagine how many normal-sized pages it would take to print a blueprint of all the circuits in a brain?

There is indeed a project that proposes something very much like that; there is talk of mapping all of these brain connections at various resolution levels in the hopes of better understanding how brains work. One of the objectives of this mapping is to be able to construct a faithful model of neural processes, eventually simulating a complete brain. Collectively, these efforts aim at obtaining a (and sometimes even called *the*) connectome[12] of a human brain. The term *connectome* was independently proposed in Dr. Patric Hagmann's doctoral dissertation and in a paper led by Dr. Olaf Sporns in 2005 and is commonly defined as the connection matrix or the total number of connections in the human brain.[13,14]

Without a doubt, the development of a connectomic map of a typical human brain would be incredibly interesting to say the least and "the mother of all breakthroughs" in my opinion, as this would be a very important advancement of our quest of truly understanding ourselves. That said, we have to think very carefully about what this achievement will mean, but even more importantly, we have to be clear on what it does not mean.

The generation of a connectome would be a major undertaking even if the technology necessary to complete this task correctly was fully optimized, which at the time of this writing, it is not, at least not as far as fine synaptic mapping is concerned. Let me clarify. There are several important aspects of brain physiology that are not explicitly accounted for in most conceptions of the connectome paradigm. It is important to realize that to map all major connections between brain areas is one thing; to map and characterize each synaptic connection is another, much more complex matter.

For example, I can get a street map of my university; this will only tell me how to get to a particular building. So far, so good and in most cases this will be good enough. However, this will tell me nothing about how things work in an individual hall (say, the electrical wiring for example). Now, if the objective is just to know how to get to a specific place, the street map will suffice. The catch is that to really understand how a university works, the street addresses or even the understanding of the electrical wiring of the buildings is not enough. All the work, all the important decisions are done mostly inside the buildings by entities not directly connected with it. Similarly, in a real brain, the internal "wiring" of the synapses is an integral determinant of function, but it is not the only important one. Therefore, it is not enough to know the particular connection map; what happens within each individual synapse (and even at a lower structural level) matters too.

Let us try to see this another way; to say that a fine map of all the connections of your brain tells you everything there is to know about you is a little like saying a picture of you tells everything there is to say about yourself. We all know, of course, that this is not so. A snapshot of a person just gives us the physical appearance at a particular point in time and from a limited perspective.

Whomever or whatever you are or I am is much, much more than the simple sum of our parts. You see, structure and sheer numbers, as we stated earlier, are not everything; the physical structures in a brain are in a state of constant change by interacting with other structures. As far as we can tell, all these dynamic interactions are what actually generate the mind. Just like life, the mind seems to be a process rather than a thing. Simply because of this fact, in addition to the structural information that may be obtained by the fine structural mapping of the brain, there is the question of how all these part interact with each other and how those multiple interactions change over time and in response to various stimuli.

The challenges with the connectome paradigm are very similar to another famous "-ome," the human genome. Many people thought of the sequencing of the human genome as an end in itself, the ultimate scientific discovery if you will. However, as herculean as the actual mapping was (and it was a great effort and a wonderful achievement, don't get me wrong), the real work began once representative genomes were sequenced. Also, there is no such thing as "the" human genome. There are many human genomes, essentially one per each human being that has ever lived, not counting identical twins, triplets, and so forth. Granted, all human genomes are very similar to each other, but distinct nonetheless. Then again, by sequencing a genome, we have merely accomplished the very first step; there are many functional aspects of genetics that go beyond knowing a gene's sequence. For example, a single gene can code for multiple versions of a protein (this is in part a consequence of a process called *alternative splicing*). Also, there is the question of how all the multiple gene products cooperate to generate human physiology. These are just two examples of all the possible issues that go beyond a simple genomic mapping. In a similar vein, all connectomelike projects must deal with similar considerations, as we will see shortly.

Of course, many structural features and connections in the brain are fixed. There is a high degree of hardwiring in the brain; otherwise, we would not be able to do anything at all (think about the street map). In this sense, the mapping of the hardwired parts is much more meaningful. But what really defines an individual person are the detailed synaptic and subsynaptic connections and interactions, the fine-mapping if you will in conjunction with the physiological phenomena in each neuron.

Here are some of the additional established neuronal mechanisms that cannot be explored or studied just by mapping brain connections, as the purely structural connectome paradigm proposes:

- In many cases, neurotransmitter *spillover* is a means of communication between neurons. In this mechanism, some of the neurotransmitters released by the synapse will end up at receptors not located at that particular synapse; obviously, in these cases, direct synaptic contact is not even necessary.
- In addition to that, there are *autoreceptors*, which are located at the same neuron where the neurotransmitter originates. In this process, a certain fraction of the neurotransmitter molecules released by a neuron will

- interact with these autoreceptors as part of a feedback mechanism, for example.
- Moreover, many synapses work with more than one neurotransmitter, a phenomenon termed, not surprisingly, *cotransmission*.
- Furthermore, in some synapses there are some substances that, while not neurotransmitters in their own right, can influence neurotransmitter physiology. We call these substances *neuromodulators* or *modulators* for short.
- There are even atypical neurotransmitters like nitric oxide and carbon monoxide whose physiology differs from the classical chemical neurotransmission principles. A major characteristic of them is that they are not stored or transported in the same way as other chemical messengers. In essence, these atypical neurotransmitters, despite not being targeted to specific postsynaptic receptors, are nonetheless able to modulate neuronal functions.
- The problem is further compounded by the fact that in many regions of the brain connections are constantly formed, eliminated, or, at the very least, changed as a consequence of normal brain function. The fact that you (hopefully) remember what I wrote three paragraphs back means that your brain is physically changing in real time; therefore, the exact number and nature of brain connections related to memory and perception are in a state of constant change. Essentially, any fine-mapping of all the connections in a brain will be obsolete as soon as the person thinks, observes something, or ponders about anything.

Another important implication about the purely structural mapping is that this would be a static connectome; essentially, a picture. Another detail is that our brain is not just composed by neurons. It also contains nonneuronal cells, collectively called *glia* or *glial cells*.[15] Glial cells are not excitable, yet they seem to be about ten times more abundant than neurons in a typical brain. This 10:1 glia-to-neuron ratio is widely reported in the general literature; however, different brain regions may display different ratios. On average, the glia-to-neuron ratio in whole human brains is more like 1:1; however, different brain regions show different ratios, with a range of about 17:1 down to 1:1 or even less. I am sure that additional work needs to be done to come up with a more accurate picture of the exact ratios, but on average the brain complexity we are talking about at the very least doubles in terms of cell numbers when we take glia into account. Further, traditionally, glial cells were thought to play just structural and support roles in the nervous system; in fact, the word *glia* means "glue."

However, glial cells are much more than glue; they are quite the active partners in brain physiology. Even though glial cells do not participate directly in neuronal signaling, they do interact with neurons and are also able to release and react to neurotransmitter molecules. In this sense, there is little doubt that glia affect many aspects of nervous system function, and therefore any true simulation or model of a brain must take into account these support cells as well.

Let's think back to the fact that a typical human brain has about 100,000,000,000 neurons, and that as we said before each of those neurons has an average of 30,000 synapses, more or less. Now think about the following: each synapse seems to be capable of memory storage and processing capabilities. Therefore, it seems that your brain or mine possesses more switches than the sum of all the computers, routers, and web connections in the world.[16]

Now, is the human brain complex enough for you?

In light of this staggering number of synaptic connections, an extension of the connectome paradigm has been proposed, entailing synaptic mapping. Not surprisingly, this approach is known as the synaptome.[17] This is certainly an improvement in the resolution for brain mapping, but it is still a purely structural approach, yet an exciting one nonetheless.

An actual example of the overwhelming nature of the problem of translating brain structure to function is illustrated by one of the most popular model organisms in modern biology, the worm *Caenorhabditis elegans*. We mentioned them in chapter 1. These little guys have around a thousand somatic (nonsex) cells, of which, exactly 302 are neurons (in a normal animal). The connectome mapping these 302 neurons and the roughly 7,000 connections between them is completed, but the interpretation of this effort in terms of behavior has proven elusive. In other words, we know the exact number and nature of all the neuronal connections of this worm, yet that is not enough to predict its behavior 100% accurately. Part of the reason for this is that knowing the structural details of the *C. elegans* neuronal connectome, although necessary, is not sufficient to account for its behavior. As we saw before when referring to the brain, we would need to consider at least some of the other additional neuronal mechanisms outlined previously.

A particularly promising approach to overcome the inherent difficulties of the "just structural" connectome approach is to create another type of mapping, namely, a "brain activity map,"[18] which is based on the assumption that any neuronal circuit function and therefore by extension brain function are emergent properties like the ones we discussed in part I of this book. This is certainly a BIG step in the right direction. One of the proposed approaches to achieve this is the determination of the activity of each neuron in a given circuit—a huge task, but a feasible one nonetheless. A variety of techniques are proposed in this effort, from recording the action potentials in all the neurons in a given brain circuit to voltage imaging to measuring the dynamics of ions in and out of neurons, as in calcium imaging. This approach will nicely complement the purely structural approach. One thing is certain: these are exciting times for brain research! This trend illustrates how the study of the nervous system can go all the way from the molecular to the systems level.

Regardless of the specific approach, one of the main objectives of connectomics is to eventually generate a computer simulation of the human brain. In this respect, there are also many ethical questions about generating any kind of brain simulation. For example, let's suppose that a brain simulation is properly constructed and this brain "wakes up"—what then? Do we have a person? Depending on the faithfulness of the simulation, I believe we would have a person indeed! If so, what happens when the experiment ends? If we turn the computer off without "saving the modified file," and if we turn the simulation on again, will we be effectively giving Alzheimer's disease to a person? Even worse, if we erase the simulation, will we be killing a person? I, for one, am ever so happy that I do not have to deal with those questions in my own research.

It is very important to point out that I am not saying that mapping the human brain in whatever form is an impossible task; I am not saying that at all. Throughout history, many a scientist has said "that's impossible!" about one thing or another only to be proved wrong sooner rather than later. I really do not want to join that group.

On that note, given the brain's mind-boggling complexity that we just saw, do we have any hope, then, to understand the human brain, not only at the structural level, but also at the functional level? Well, actually, yes, there is more than hope; we have two very important possible strategies, which are not mutually exclusive, that we can use. First, we can apply the reductionist approach that we saw in chapter 1 by breaking the brain into manageable parts. In this way, we can facilitate the study of the brain, keeping in mind, as we discussed, that whatever the brain does is, indeed, more than the mere sum of its parts. This is the basic strategy used in the various connectome projects. Also, we can use simpler animal models that can give us an initial understanding of everything that a human brain entails by exploring animals' simpler nervous systems. This is an approach that we will explore in more detail soon. Before that, let's explore some general aspects of a discipline that is intimately related to the neurosciences, the field of pharmacology.

Corpora non agunt nisi fixata.
(Entities cannot interact unless in contact with each other.)
—PAUL ELRICH

5

SOME BRIEF THOUGHTS ON PHARMACOLOGY

WHAT IS PHARMACOLOGY?

ONE OF THE various definitions of pharmacology is the study of drugs and their effect on people and animals. Of course, there are many possible ways to study any given drug, and not surprisingly, this means that pharmacology can be divided into distinct yet closely related subdisciplines. The integration of these disciplines is an example of what we saw in part I when exploring the several levels of organization at which we can examine a living system. In other words, we can explore pharmacology from various points of view, namely, from the molecular level all the way to behavior.

The main goal of pharmacology is the discovery of substances that may be of use to treat or prevent disease and to alleviate conditions like pain, for example. Other relevant aspects of pharmacology include sociological and anthropological aspects, as in the case of the problem of drug abuse and the origin of pharmacotherapies, which, interestingly, can trace its origins to a lesser-known field of study called *zoopharmacognosy*, which is the study of how certain animals can self-medicate.[1,2] This can be surprising to most people, but it is well documented that a wide variety of animals consume specific plants or other material like clay as a behavioral response to sickness. Such animals include wild animals like elephants, deer, and even primates, among others. Domestic animals also practice zoopharmacognosy; for example, if you ever saw a dog or a cat chewing grass in the yard you probably saw zoopharmacognosy in action. Many of the plants that animals use for self-medication are now known to possess antimicrobial, antiparasitic, and psychoactive actions, among other properties.

As with all other biomedical sciences, pharmacology has an essentially interdisciplinary scope. An important discipline is pharmacology's sister, toxicology, which is the study of the harmful effects of drugs and other chemicals on biological systems. The lines between pharmacology and toxicology are oftentimes not well defined because any chemical that goes into an organism can be toxic or therapeutic depending on the amount. Even the term *amount* is relative, because this will depend on the specific nature of the substance, as well as on the particular physiology of the target organism and on specific environmental conditions. Toxicology has a somewhat wider scope than pharmacology because toxicology also deals with the exposure to substances that we would not usually think of as drugs, like heavy metals, for example, including lead, mercury, and so on.

A thorough exploration of pharmacology is well beyond the scope of this book. However, it is important to talk about a series of concepts that are essential in order to better understand pharmacology in general and pharmacology in light of the use of animal models in research in particular.

It is self-evident that the useful or deleterious properties of a chemical compound determine whether we are talking about a drug or a toxic substance. Therefore, the most fundamental property of such compounds is their effect on biological systems as expressed by a change in an organism's physiology. In turn, such physiological changes are in great part a reflection of three key types of interactions between a given compound and a variety of molecular targets. These interactions are expressed in terms of the concepts of *binding, receptor,* and *affinity*.

The binding concept was explicitly expressed for the first time in 1913 by Paul Elrich, one of pharmacology's pioneers, with the Latin phrase "Corpora non agunt nisi fixata," which can be translated as "Entities cannot interact unless in contact with each other." Personally, I like an alternate translation (by yours truly) a little better: "Everything starts with binding." This makes sense. If we think about it, for anything to interact with something else, there must be a connection between the two, usually in the form of some kind of physical interaction.

This brings us to the receptor concept. A few years before Elrich, another pharmacology pioneer, John Langley, proposed in 1905 that drugs specifically interact with "receptive substances." We know them today simply as "receptors." Interestingly, even though the receptor concept was widely accepted and fully embraced by the scientific community (after all, it does work; it is consistent with the theory and with the experimental data), the very existence of receptors was not conclusively demonstrated until the 1960s.

How does this receptor business work? Briefly, substances very much like the neurotransmitters that we saw in chapter 3 interact with specific proteins to activate and control a wide variety of physiological activities. Such activities include but are not limited to the opening or closing of ion channels, muscle contraction, and cell signaling, among many other processes. At the molecular level this entails the physical contact between these substances at their receptor targets. In general, these signaling substances, like neurotransmitters, for example, are termed *ligands*.

Any ligand–receptor interaction is controlled in great part by the complementary molecular shapes of the ligand and the shape of a specific region of a receptor; we call this region of the receptor its *binding pocket* or *active site*. The relationship between the ligand and its binding pocket is similar to the relationship between a key and a keyhole. Only those ligands that have the right size and shape will be able to activate their receptor target. Therefore, you can think of the binding and activation processes as the interplay between a key and its corresponding lock. If you want to open the door, or in other words, if you want to elicit a certain response from a receptor, you need the right key (the ligand) bound to that door's lock (the receptor's binding pocket).

Strictly speaking, several other factors besides shape influence the binding of the ligand to a receptor and the induction of a specific action. These include temperature, acidity, and some other environmental conditions. However, the "right" shapes and correct spatial orientations must be there, no compromise.

Before we go on, it is worth taking a brief digression to clarify a misconception about what we mean by "binding" in the pharmacological sense. In colloquial conversations, the term *binding* implies that something is fixed in place, immutable, as in the phrase *binding contract*.

However, when we talk about "binding" in a pharmacological context, we are talking about a rather dynamic process as opposed to a static one. To visualize this, we must keep in mind that at the microscopic realm, atoms and molecules are not little marbles frozen in place—anything but! All matter vibrates, in great part as a consequence of temperature; in fact, temperature *is* defined in terms of molecular movement or vibrations. The concept of *absolute zero*, in other words, the lowest possible temperature, bar none, means the total absence of molecular movement. That's precisely why it is called absolute zero; when molecules are absolutely still, they cannot be any "stiller."

As we mentioned before, a ligand is only able to bind to its target because of the multiple molecular features between it and the receptor alongside other factors, mostly environmental. We have a term for the "willingness" of a ligand to bind to its target; we call it the *association rate*. When a ligand finds its receptor target, it stays attached (bound) to it for a certain defined period of time; this time period can be very brief, in the range of microseconds or even shorter (a microsecond is a millionth of a second). In time, the ligand detaches from the receptor, which brings us to another important concept, the *dissociation rate*, which is the "willingness" of a ligand to detach itself from its receptor.

It is important to keep in mind that as with many things in nature, the relationship between a ligand and a receptor has a statistical aspect to it. What I mean by this is that in the real world, there will be a huge amount of both receptors and ligands. Therefore, at any given time there are multiple association and dissociation events. The amount of time that a particular type of ligand remains attached to a receptor slightly varies from molecule to molecule. For example, a certain receptor–ligand pair will remain physically connected to each other for a microsecond, while another pair will remain associated for 1.1 microseconds, a third pair remains bound for 0.9 microseconds, and so on. In a typical binding experiment what we essentially get is an ensemble of the millions upon millions of such pairings, which will average to about 1 microsecond.

Still with me?

So far we have established that at the molecular level, the interplay between ligands and receptors is essentially the continuous association (attachment) and dissociation (detachment) events that occur between these two types of molecules. Now, recall that, as we said earlier, in nature there is not just one ligand molecule or one receptor target; there are millions, even billions (or more) of each. Under these conditions, there will be multiple association and dissociation events happening at any given time. A consequence of this is that in time, we can reach something called an "equilibrium state" for a specific ligand–receptor pair population. At this equilibrium state, the number of dissociation events is roughly equal to the number of association events. It is only then that we can say that a ligand is "bound" to its receptor in a pharmacological sense. Once bound, the main consequence of the binding of a ligand to a receptor is to increase the statistical probability of receptor activation; therefore, we can consider the population of receptors "active" if a significant number of them are active at a certain point in time. Of course, sometimes things can get more complex; we can talk about "initial binding rate" and similar terms, but we do not need to get into that here.

The third important concept in pharmacology we will mention here is the concept of *affinity*. As the name suggests, affinity gives us an idea of how much a ligand "likes" being bound (again, in the pharmacological sense) to its receptor target. Therefore, we can describe affinity in light of the relative association and dissociation events in a system. Briefly, if we have two different types of ligands that bind to the same receptor pocket, the ligand with higher affinity associates to the receptor faster that it dissociates from it. Conversely, a ligand that displays lower affinity dissociates from the receptor faster than it associates to it. In physical terms, a ligand with high affinity spends more time (on average) attached to its receptor target than detached to it.

Up to this moment, we have described drug effects at the microscopic level. This way of describing the effect of chemicals in a biological system is broadly termed *pharmacodynamics*, which is the study of drug targets and the ligands that associate to them, as well as the physiological consequences of such binding. These receptor targets are generally, though not exclusively, proteins, and as we saw earlier, proteins are built based on the specific genetic endowment of the organism.

In contrast, at the whole organism level, the effect of any pharmacological agent is described in terms of its *pharmacokinetics*, which is simply the path a drug takes in the body in terms of time and its association with four main parameters, *administration, distribution, metabolism,* and *excretion*, which together are known as *ADME*. All drugs have their own ADME scheme. There are various ways to *administer* a drug; we are all familiar with the usual suspects (pills, injections, ointments, etc.). Once in the body, drugs are *distributed* through the organism in various ways, for example, via the circulatory system. At the molecular level, drugs interact with at least one target, although in reality, most drugs interact with multiple targets; this partly accounts for the infamous side effects. Drugs are also *metabolized* in the body; this metabolism usually implies changes in the chemical composition of the drug, which can result in the inactivation of the drug, which can enhance its *excretion* via the usual ways (sweat, urine, feces, etc.).

We can use all these concepts (association, dissociation, affinity, ADME, etc.) to describe practically any kind of drug (or toxin)–receptor interaction. Of course, all these molecular events may lead to minor or major changes in the organism. In this case, we can talk about the efficacy of a certain drug or the toxicity of others. This is pretty much what pharmacology is about, and the understanding of these concepts is absolutely essential to make sense of how medications induce their effects. There will be variations of the theme, of course, but in general these concepts apply to any system that we want to study in the pharmacological sense. Now let's talk about some of nature's master pharmacologists, plants.

PSYCHOPHARMACOLOGY, PLANT STYLE

Psychoactive substances are chemicals capable of inducing changes in mood, perception, and related subjective states in animals. It is widely known that many species of animals consume plants that induce psychoactive effects. That said, when we think about it, this is rather strange; why would a plant synthesize psychoactive substances? Plants do not seem to need psychoactive substances for anything, as these compounds do not seem to play any metabolic role in the plants that make them. Moreover, many of these compounds are metabolically expensive; therefore, it does not make sense for a plant to invest valuable metabolic resources to synthesize a compound that does not appear to be essential for its survival. Also, as we discussed, plants do not have a "traditional" nervous system, and as far as we know, a nervous system is necessary to produce the phenomenon of "mind" (but please remember the section on plant neurobiology in chapter 3!). Seen in this light, again, plants do not seem to display any "psycho" aspect that may be affected by psychoactive substances. What's going on? In other words, how does the production of psychoactive substances benefit plants?

Briefly stated, evolution, pure and simple! I am convinced of this; and even if I were not convinced, it wouldn't matter, as the scientific evidence is undeniable. As part of this process, chemical structures that prove beneficial for the survival of the organism are conserved. These molecules, among other things, may act as toxins against macromolecules such as proteins, which control important physiological processes in living organisms, as we saw in chapter 2. Alternatively, some of these substances may serve to attract insects, for example, with an eventual benefit for the plant—for example, pollination (a common theme in biology is to maximize the chances of reproduction). The study of the interaction of chemicals and living organisms is an aspect of the discipline of *chemical ecology*.[3]

The study of chemical ecology is intensively pursued, and rightly so; there is abundant evidence suggesting that plants use chemicals as a defense from insects and other pests. This can have consequences beyond the mere description of the phenomenon. There can be implications and applications related to important human activities such as agriculture and the discovery of medications. One of the first and still best examples of this is that the field of local anesthesia was started by a natural compound thought to have a defensive role that helps its originating plant, but I am getting a little ahead of myself here. Let's continue talking about some aspects of chemical ecology.

Many of these defensive substances produced by plants are classified as *alkaloids*. I start with alkaloids because I want to clarify right away a very important point. Sometimes people utter the word *alkaloid* with almost reverence, as if the term implies something deeply mysterious. That is not so. The term *alkaloid* merely means that the molecular structure of a certain compound includes a nitrogen atom bearing a specific relationship to other atoms in the molecule, often as part of structures called *heterocyclic rings*. Heterocyclic rings are simply

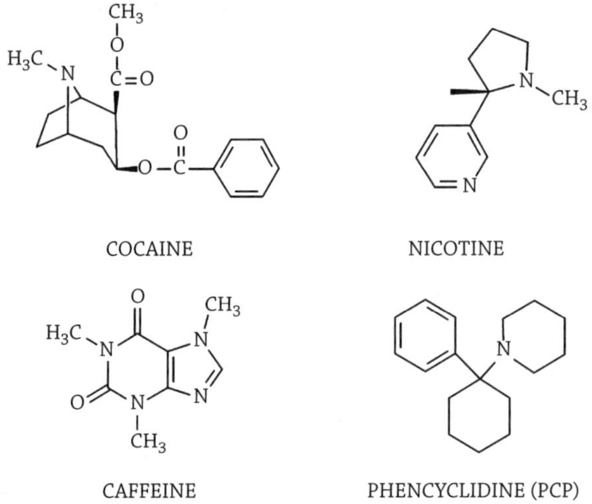

FIGURE 5.1. Representative alkaloids. These alkaloids also happen to be abused drugs. Illustration by O. R. Pagán.

molecular rings that contain at least two different kinds of atoms in a molecule. In most cases the ring is mainly composed of carbon atoms with one or more nitrogen atoms completing the ring (Figure 5.1). Many types of alkaloids include well-known psychoactive drugs and abused substances as shown in Figure 5.1.

Frequently, substances like alkaloids are termed *secondary metabolites*, with the unfortunate consequence that the *secondary* term gives the wrong impression that these compounds are of lesser importance. This is yet another example that illustrates that terms matter! Please recall the question about "basic" versus "fundamental" research in chapter 2. This is just another example; again, the word *secondary* as used in *secondary metabolites* can give the impression that such compounds are less important—nothing further from the truth! In this context, what it means is that secondary metabolites are compounds that are present only in certain types of plants, as opposed to *primary* metabolites, which are compounds that are present at a wider scale.

For example, all photosynthesizing plants have at least one version of *chlorophyll*, which allows the plant to capture light, harvest its energy, and transform it into chemical energy. In this sense, chlorophyll is a *primary* metabolite by its virtue of being present in all green plants. On the other hand, not all plants synthesize caffeine (ah, one of my favorite plant-derived substances; see Figure 5.1); just a few species do. Therefore, caffeine is considered a *secondary* metabolite. Many abused drugs of natural origin are secondary metabolites, including the well-known compound *nicotine*, produced primarily by tobacco plants; incidentally, nicotine (Figure 5.1) is probably one of the most addictive drugs used by humans.

Nicotine, as well as other abused drugs like cocaine, has been described as natural insecticides in the sense that oftentimes they fatally interfere with the normal physiology of an insect's nervous system. Insects are much smaller than humans; therefore, an insect that feeds on a tobacco plant is exposed to nicotine, and this exposure will repel the insect and can even kill it (think overdose). On the other hand, a human (a much larger organism) exposed to the same amount of nicotine can experience psychoactive effects. Why? The reason, again, has to do with evolution. In the same way that many defensive molecules are conserved for their positive effects on survival, their molecular targets are conserved as well. In other words, vertebrates, including humans, will express targets very similar to the ones present in insects due to our shared evolutionary history. In this example, a human that smokes or chews tobacco is exposed to nicotine, which will then interact with these targets, and since humans are much larger than insects, instead of being a toxin or even an irritant, nicotine can induce pleasurable sensations. The similarity between many insect and vertebrate molecular targets is part of the reason why invertebrate animal models have provided important insights about human physiology. It is important to point out, though, that any substance, nicotine included, can be toxic to humans if taken in high amounts, as we said at the beginning of this chapter. Nicotine's principal targets are ligand-gated ion channels called nicotinic acetylcholine receptors, an important type of receptor that we'll see in somewhat more detail in the next chapter.

As we saw, another plant-produced psychoactive substance is the alkaloid *cocaine* (see Figure 5.1). We are all familiar with the drug abuse aspects of cocaine consumption, but before these negative effects were discovered, the original use of cocaine was as a *local anesthetic*. Cocaine was the very first example of a naturally occurring molecule with local anesthetic properties. Cocaine was precisely the substance that actually initiated the field of local anesthesia.

Speaking of local anesthesia, before we talk about other pharmacological aspects of cocaine, we must first take a brief detour to your favorite dentist's office. (Wait a minute, is it even *possible* to have a favorite dentist? But I digress....)

The development of anesthetic agents was a blessing for humanity. In my opinion, there is no better way of saying it. Close to two hundred years ago, the only anesthetic agents widely available were several big, strong guys who held the patient still and in place while the surgeon worked as quickly as possible. This was also true, albeit less dramatically, in dental procedures. Interestingly, in those times dental interventions were usually performed by the town's barber, who also served as the town's dentist. Moreover, not surprisingly, dentists, rather than physicians, were the first to use anesthetic agents, therefore initiating the general field of anesthesia. Nowadays, a wide variety of anesthetic compounds are a standard, even essential component of dental and medical practice.

As we said, the first example of a compound showing local anesthetic properties was cocaine. Cocaine's anesthetic effects were evident well before its addictive and toxic effects were recognized. The history of the discovery of cocaine is an excellent example of how a naturally occurring substance finds its way into medical applications. Cocaine is found in close to two hundred plant species of the genus *Erythroxylum* (this term means reddish). These plants are widely distributed in parts of Central and South America and, oddly, in Madagascar, on the other side of the world. South American natives named these plants *khoka*, which sounded like "coca" to the Spanish Conquistadores. In Quechua, one of South America's native languages, khoka means "the plant"; this is a strong indication of the importance given to this plant by this Native American culture. Most historians agree that coca leaves were used as early as about 5,000 years ago. South American natives chewed on them to take advantage of their mild stimulant effects and the human dependence on coca leaves was described by European colonizers.

It seems that the first written report that mentions the anesthetic properties of coca leaves was published in 1653. The author of this report was a Spanish priest, Bernabé Cobo, who was also an avid amateur naturalist. Apparently, the story began when Cobo went to his barber to take care of a toothache. The barber told him that he'd rather not pull the tooth out since it seemed healthy enough. So the good Father Cobo, on the advice of a fellow *padre*, started chewing on coca leaves.[4] Lo and behold, pretty soon the pain subsided! History does not record whether Father Cobo kept chewing on coca leaves after his toothache was gone.

The chemical isolation of cocaine from coca leaves was first achieved by the German chemist Albert Niemann in 1860. Niemann was a disciple of the famed organic chemist Friedrich Wöhler, who is famous for his many chemical successes, especially for synthesizing urea for the first time completely in the laboratory, the story used to begin most organic chemistry textbooks. The isolation of cocaine was the subject of Niemann's PhD dissertation titled "On a New Organic Base in the Coca Leaves."[5] In his dissertation, Niemann actually coined the term *cocaine*, in an analogy with other alkaloids like nicotine, morphine, and so forth.

Oddly, even though cocaine's anesthetic properties were well known and noted by various authors at the time, it was not until twenty-odd years later, in 1884, that cocaine was tested as an anesthetic agent in ophthalmology. Ironically, one of the first uses of cocaine was to try to alleviate morphine addiction! This was proposed by no other than Dr. Sigmund Freud, who by the way was a neurologist, not a psychologist (psychology as a profession was not yet developed in Freud's time). Freud also has the dubious distinction of being the first "famous person" to become addicted to cocaine.

The use of cocaine as a local anesthetic was started by Dr. Carl Koller, a friend and colleague of Freud's. They experimented with cocaine quite extensively on themselves, and this is no euphemism. In one of their sessions, Koller put some cocaine on his lips and noticed that his lips went numb. He mentioned this to Freud, who noticed it too, but thought no more of it (or maybe he did not care). Anyway, Koller recognized the importance of this and later ran a series of animal experiments that conclusively demonstrated the anesthetic properties of cocaine, particularly its promise as an ophthalmological anesthetic. Even today cocaine is still used, albeit in a limited way, for certain types of surgeries, usually minor surgeries from the neck up, because in addition to being an anesthetic, cocaine also induces vasoconstriction, which limits excessive bleeding.

The worst possible effect of cocaine abuse is sudden death, which is usually attributed to the induction of cardiac arrhythmias. Paradoxically, this effect is not necessarily related to addictive behavior, since a first-time user can die from an overdose. At subtoxic concentrations, cocaine can affect virtually all organ systems, with particularly evident effects on the nervous system as a powerful locomotor stimulant and mood enhancer.

At the behavioral level the accepted target for cocaine is the monoamine transporter (MAT) superfamily, which includes the dopamine transporter (DAT), serotonin transporter (SERT), and norepinephrine transporter (NET), among others. These and related proteins control the reuptake of neurotransmitter molecules in chemical synapses. This is one of the mechanisms of neurotransmitter inactivation that we referred to in chapter 3. Neurotransmitter reuptake by specialized transporters was proposed as such an inactivation mechanism about fifty years ago.[6] Upon blocking of the dopamine transporter by cocaine, the concentration of dopamine increases in the synaptic gap, which causes hyperactivation of postsynaptic dopamine receptors, as well as indirectly activating extrasynaptic dopamine autoreceptors. This abnormal increase in the neurotransmitter concentration seems to account for the behavioral effects of cocaine and other compounds.

For some time, the only source of cocaine was the coca plant, as the synthesis of cocaine proved elusive, since its chemical formula can correspond to eight different cocainelike structures (the wonders of organic chemistry). Of these, only one is the "correct" one in the sense that it induces the anesthetic and behavioral effects associated with cocaine intake or administration. The chemical synthesis and the elucidation of the cocaine structure were achieved by the German chemist Richard M. Willstätter in the late 1800s.

Once the use of cocaine became widespread (yes, a famous soft drink, a type of wine, cough drops, and even children's toothache drops used to contain cocaine), its toxicity and addictive properties became immediately apparent. Cocaine addiction was eventually called the third scourge of humanity (the first two were alcohol and morphine addiction), and its use as an over-the-counter drug was discontinued in the early twentieth century.

Now, going back to the question of why psychoactive substances are beneficial to the plants that produce them, a possible interpretation is that many types of plants are engaged in a "chemopsychological war" against animals.[7] Therefore, plants that produce such substances may have an evolutionary edge at several levels. The first one, as mentioned before, is the immediate protective effect against insects. A second example is more subtle. Humans, upon experiencing pleasurable sensations from the use of a given plant species, will start a *selection process*. In this process, plants with desirable properties are protected from insects or grazing animals and eventually actively cultivated. This can also happen when other particular properties of the plant are discovered, for example, anti-inflammatory properties or pain alleviation, as illustrated by the coca plant. Eventually, a population of plants with the desired traits is obtained. With time, further selection of plants showing more powerful effects can happen. This is a time-honored practice still widely performed in agriculture and animal husbandry today. In fact, this *artificial selection* inspired Charles Darwin, who went on to formulate the concept of *natural selection* (see chapter 1).

Nowadays it is pretty much accepted that these human-initiated processes, like artificial selection, are major players in the game of evolutionary change in nature. Some schools of thought even propose that plants that produced psychotropic substances coevolved with humans.[8] For example, nicotine mimics the physiological effects of the neurotransmitter *acetylcholine* (see chapter 3, Figure 3.5) of central importance in the nervous and muscular systems in vertebrates, including humans.[9] Other compounds like cocaine and morphine modulate the physiological responses to dopamine and naturally-occurring opiates (endorphins), respectively, which are important neurotransmitters in their own right. Based on these and other examples, the current thinking is that the ingestion of neurotransmitterlike molecules derived from plants by early humans acted as substitutes for bioenergetically expensive neurotransmitters under nutritional deprivation states. In other words, we can say that these plants acted as the first nutritional supplements ever used by humans.

Not surprisingly, human–plant coevolution has sociological implications. Traditionally, people who acquired detailed knowledge about the effect of plants in humans were considered sacred persons, also known as holy men (and women) or shamans. This practice has apparently been around for quite a long time. In the late 1950s archeologists discovered and explored an ancient site called *Shanidar Cave* in today's Iraq, where several Neanderthal remains were found. One of them was Shanidar IV, a Neanderthal male who appeared to be buried in a ritual manner. In Shanidar IV's burial site, researchers found many types of plant pollen, most of them later identified as coming from plants that are recognized today as having medicinal properties. This data was interpreted as evidence for considering Shanidar IV's remains as those of a medicine man, who was buried with the plants he worked with so he could use them in the afterlife. This is not as far-fetched as it may sound. Many more recent cultures have engaged in similar practices. That said, there is still controversy about considering Shanidar IV as a medicine man, but if true, this would represent the earliest example of the medicinal use of plants by humans (yes, Neanderthals were humans, but this is a story for some other time).

This story brings to mind a very important question: how were psychoactive plants discovered anyway? It is safe to assume that we will probably never know for sure; this happened thousands of years ago, before history began to be recorded, as we saw earlier. Also, it is very possible that the story will be slightly different depending on the specific substance. Nonetheless, we can have some fun imagining a few plausible scenarios. It is worth pointing out that these scenarios also apply to plants that induced effects not exactly psychotropic, like the ones with antiparasitic or antipyretic (antifever) effects, for example. One possible scenario would be when a person ate a fruit or some other part of a plant. If the plant produced a substance capable of inducing a psychoactive effect, it would be apparent in a very short time.

A second possibility is if someone took a stroll through a field and nibbled on a blade of grass; most of us have done that at one point or another. The person may have noticed a peculiar flavor or a numbing sensation in the mouth. I think that this was the most likely way by which coca leaves were found to have anesthetic properties. As we have seen, we know that South American natives were aware of this, as they used to chew on coca leaves to alleviate toothaches and for other purposes. This was the most likely source of the information that was relayed to Father Cobo.

Finally, an amusing yet entirely logical possibility: let's suppose that after the discovery of fire, a group of humans needed some kindling (like dry leaves) to start a campfire and all they found were some dry marijuana leaves and flowers (I just know that you know where this is going). Anyway, they started the fire, gathered around it to warm themselves up, inhaled the smoke, and, well…the rest, as they say, is history.

ANIMAL MODELS IN PHARMACOLOGY

In chapter 2 we talked about the various choices that we have available to apply to biomedical research. These alternatives go from *in vitro* studies all the way to *in vivo* experiments, which imply the use of animal models. Just to remind you, what determines the relative usefulness of any particular model depends on the actual research to be done, along with the specific characteristics of the experimental organism. Many models are better suited to examine complex behaviors, like vertebrates, for example, especially mammals. Others may have the advantage of being amenable to molecular biology due to the availability of their sequenced genome, and some others may possess some peculiar characteristic, like unusually large neurons, for example (think the giant axon of squids). Specifically in genetics, several animal models were chosen based on their relatively short generation time; in other words, they reproduce fast, which is a distinct advantage when working on genetics research.

When dealing with pharmacology, the difficulty in choosing an adequate animal model is amplified because in general, pharmacological responses are varied, even within the same species. This variation with regard to the response to drugs is, of course, largely a consequence of genetics. In general, a given organism tolerates genetic variations that affect pharmacological responses because the genes that control the response to drugs are more resistant to mutations than other genes that code for processes essential for the life of the organism. Let's explore this a little further.

There are many aspects of physiology that are absolutely essential for the survival of the organism. One of those is the capacity to harvest energy, either from ingesting other organisms or directly from the environment, as in photosynthesis. If we do a survey of the genes that deal with these types of processes, we will find very little genetic variability, or at the very least, relatively less variation compared to other types of genes. This is not surprising, since if the ability of an organism to capture energy is not negotiable in terms of survival; if this ability is compromised by a genetic mutation, so is survival and therefore reproduction. In other words, when speaking about these fundamental processes, there is very little wiggle room, genetically speaking.

On the other hand, many pharmacological responses are more resistant to variation and therefore can be present in multiple forms in a population. A very common example in humans is alcohol metabolism. We all know people who are very good at "holding their liquor," and we also know people who can get dizzy with one beer. Of course, we also see everything in between. The way you react to alcohol is a reflection of your individual genome. In this sense, the genes (yes, usually there is more than one) that affect the physiology of drug metabolism will tend to be more tolerant to mutations. In other words, if you have a mutation that makes you more sensitive to alcohol, you can adapt your behavior, for example, by limiting or even eliminating alcohol intake. If you do this, you are in essence rendering that mutation irrelevant. On the other hand, as we saw earlier, most mutations in genes controlling the physiology of oxygen distribution in an organism, for example, will most likely tend to be harmful. An example of this is the case of sickle cell anemia, a disease in which a mutation in a gene that codes for a hemoglobin chain produces an abnormal protein that limits the ability of red blood cells to carry oxygen to tissues, with the expected harmful effects.

The net result of this in terms of pharmacological responses is that since mutations may not matter that much in a manner of speaking, this naturally allows for multiple versions of genes related to drug metabolism. Incidentally, this does not only apply to alcohol. This applies to practically every single medication or abused drug and essentially any exogenous chemical substance that can be ingested by an organism.

A related factor involved in the metabolism and effect of any exogenous compound intended for clinical use is the specific genotype of the patient. A relatively recent development in pharmacology is the rise of pharmacogenetics, sometimes called pharmacogenomics. In a strict sense, the two terms are different. Pharmacogenomics deals with the study of the various genes that determine the drug response. On the other hand, pharmacogenetics deals with the study of genetic variation on drug response. As you see, there is a subtle distinction, but again, the two terms are often used interchangeably. One of the consequences of genetic variation, as we have seen before, is the differential response to drugs in different individuals.

These are some of the aspects that make pharmacology research so interesting but at the same time so challenging, especially, again, when needing to choose an ideal animal model.

As you can imagine, the study of pharmacology in relatively complex animals is made more complicated by these factors. One approach is to study isolated cells, as in the *ex vivo* approach we saw in chapter 2. This, of course, has some fundamental limitations because to really observe, let alone understand, pharmacology, we must look at complete physiological systems. In this respect, the study of whole organisms is essential. That said, simplicity is an advantage, too. In essence, the choice of experimental organism is tricky. This is especially true in experimental pharmacology because an optimal experimental organism for these purposes needs to be simple enough so that its analysis is tractable while displaying the behavioral traits that need to be studied. In the specific case of behavioral pharmacology, the capacity of showing varied behaviors is particularly desirable. An emerging class of organisms that meets these standards includes some of the most interesting types of life forms on earth, the flatworms, which we will talk about next. These little wormies will be the main characters in the rest of this book.

III

Planarians

> *Planarians are rather wonderful and challenging beasts whose rather special biological and psychological talents and properties can make them fascinating subjects for scientific investigation. The worms will provide the fascination—it is up to you to provide the science.*
> —JAMES V. McCONNELL

6

PLANARIANS

WHAT IS A FLATWORM?

BEFORE GETTING INTO what a flatworm actually is, we should explore a more general question: what is a worm? A "generic" worm is an invertebrate, which means an animal that does not have a vertebrate-style spinal cord. Examples of invertebrates include insects, arachnids, and, well, worms, among others. Worms in particular are usually not much more than a tubelike-shaped critter, which generally lacks appendages. There are variations of the theme of course. For example, sometimes we refer to caterpillars (the very hungry kind or otherwise) as "worms," and there is a particularly interesting type of worm, the *velvet worm*, also known as *Peripatus*, which is not really a worm but is nonetheless a very interesting little guy who I think deserves a book of its own, but I digress.

In general, worms tend to be slimy little living things. All kids like worms: girls or boys, it does not matter. The "traditional" aversion to slimy, wriggly living beings comes later in a kid's life depending on his or her particular upbringing. Some people outgrow their fascination with critters like these; thankfully, I never did. Of all worm species, flatworms are some of the most interesting ones on this earth of ours.

In chapter 2, we briefly talked about the need to classify living organisms to understand them and their place in nature a little better. In the early period of taxonomic classification, all wormlike animals belonged to the class *Vermes*, which is Latin for, well, "worm." In 1864 Ernst Ehlers defined organisms belonging to

96   Planarians

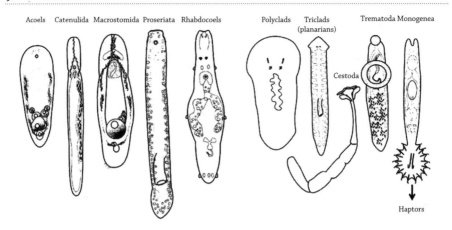

FIGURE 6.1. Representative flatworm types (not drawn to scale; their size range is from the mm range to several centimeters long).
Adapted from Tyler S (1999) Systematics of the flatworms-Libbie Hyman's influence on current views of the Platyhelminthes. In: Libbie Henrietta Hyman: Life and Contributions. Edited by Winston JE. Novitates 3277:1–66. Courtesy of the American Museum of Natural History; reprinted with permission.

the class *Vermes* as "...bilaterians without rigid skeletons, but with tubular reinforcement between epidermis and body-wall musculature, all of which function together as the main system active in the locomotion of vermiform organisms."[1]

In other words, worms are living tubes that crawl.

The term *flatworm* refers to a wide variety of organisms. In modern taxonomy, though, most flatworms belong to the phylum *Platyhelminthes*, which literally means (you guessed it!) "flatworms." This phylum is one of the largest ones in the animal kingdom; however, the exact number of species is unknown. Some estimates propose anywhere from 50,000 to 100,000 representative species (Figure 6.1). One thing is for sure, there are a lot of flatworms out there!

A detailed description of the formal classification of flatworms is beyond the scope of this book; nonetheless, let's do a brief overview of the major criteria to organize this varied group of animals. Traditionally, the classification scheme for the *Platyhelminthes* includes four main groups. Three of these groups include parasitic organisms, including the *Trematoda* (example: liver flukes), *Cestoda* (example: tapeworms), and a relatively lesser-known group of parasites, the *Monogenea*, which are mainly worms that externally attach themselves to fish through a curious structure that can be likened to a combination of suckers and hooks. These structures have a rather curious, "science fiction-y" name: *haptors* (Figure 6.1).

All of these parasitic flatworms are fascinating in their own right. Besides their interesting biology, the study of flatworms is also of practical importance

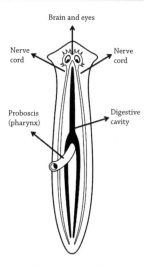

FIGURE 6.2. Simplified general anatomy of a typical triclad. Illustration by A. G. Pagán.

due to the millions of people in the world affected by them. However, the study or parasitic flatworms is outside the scope of this book.

There is a fourth group of platyhelminths, the *Turbellarians*, which are free-living (nonparasitic) flatworms that are widely distributed in nature and compose about 16% of all flatworm species. Figure 6.1 also shows two of the best-known types of turbellarians, the polyclads and the triclads. The term *turbellarian* was coined by C. G. Ehrenberg in 1831, based on the type of locomotion that most of these organisms use. In turbellarians, ventral cilia (fine hairs) beat in a coordinated manner, causing water turbulence.

Turbellarians are widespread in nature and were originally classified based on their habitat by Paul Hallez in 1892. For example, there are turbellarians that live in fresh water (group *Paludicola*), some that live in the ocean (group *Maricola*), and even many terrestrial species as well (group *Terricola*). These terrestrial species generally feed on other types of worms, mainly earthworms, but all turbellarians are predators, scavengers, or both. Broadly speaking, the turbellarians are divided (nontaxonomically and kind of arbitrarily) according to the size of their adult form. Briefly, microturbellarians are about 1 mm long or less; an example of these is the marine worm *Macrostomum lignano*, of which we'll talk about later. On the other hand, macroturbellarians are longer than 1 mm. The only two types of macroturbellarians are the *triclads* and the *polyclads*. The triclads are the ones that most people know as *planarians*; they display three main branches to their digestive system, one of which points toward the head and the other two toward the tail, forming an inverted "Y-like" structure (Figure 6.2). On the other hand, the polyclads display a central digestive cavity branched evenly throughout the animal's body.

The record for the biggest freshwater flatworm seems to belong to *Rimacephalus arecepta*, a triclad species that lives in Lake Baikal, Russia. This worm can typically grow up to 90 mm long (about 3½ inches),[2] but there are reports of individuals reaching about 400 mm long (about 16 inches).[3] Terrestrial flatworms tend to be even bigger; some species can be about 800 mm long (about 31 inches!).[4]

The polyclads are a particularly interesting class of flatworms. Several polyclad species include some of the most colorful organisms in the sea. As with many brightly colored organisms, their coloration seems to serve as a warning to other organisms, as many polyclads are poisonous. The polyclads also seem to have the most complex nervous system among the flatworms; these nervous systems show some very intriguing traits. In a series of curious experiments that would not be entirely out of place in a horror movie,[5,6] brains from a polyclad species (*Notoplana acticola*) were transplanted to other polyclads of the same species, and the transplants held! When transplants "hold," it means that the recipient of the new brain did not reject it immunologically. Surprisingly, after a recovery period of about 48 hours, the worms regained almost normal behavior. Let's say this again in other words; researchers took the brains of some worms and successfully "installed" them into other worms, again, with full functional recovery.

It gets better (or worse, you decide). In follow-up experiments, polyclads' brains were dissected off and put back in the same worms in some experiments or in other worms in other experiments. There was a catch, though; the brain was transplanted at a 180-degree angle. In other words, researchers put back the brains *backwards, pointing at the tail section*. Surprisingly, as before, the worms recovered proper motility. Furthermore, researchers did experiments in which they took the brains out of the organisms' heads and put them back more than halfway down their bodies. You guessed it; they recovered almost normal behavior.

The curious thing about polyclads is that they are not able to regenerate their nervous systems "from scratch" in contrast with other turbellarians, as we will see in further chapters. However, if the polyclad brain is left relatively intact during transplant, as evidenced by these results, the worm's brain is masterfully capable of finding the right connections. Flatworm brains are fascinating. We will talk about flatworm nervous systems in more detail in chapter 9.

Other researchers studied developmental biology and regeneration using freshwater planarians by doing tissue transplants between animals of the same species like *Dugesia japonica*[7] and *Dendrocoelum lacteum*[8] but also between closely related species like *Girardia tigrina*, also known as *Dugesia tigrina,* and *Girardia dorotocephala*, also known as *Dugesia dorotocephala*.[9] The curious thing is that in many instances these transplants between two different species held. This is a curious phenomenon. Please recall the example of the cheetah that we talked about in chapter 2. In "higher" organisms, unless the donor and the recipient are very similar in genetic terms, similar types of transplants (always within the same species and never involving nervous tissue) do not hold. These capacities seem to be related to the regenerative properties of planarians. We'll talk some more on regeneration and planarians in the next chapter.

As we said before, not all flatworms are formally classified as platyhelminths though. Taxonomy is a fluid science in the sense that is fine-tuned, reinterpreted, modified, and even rethought as more information becomes available. For example, there are two types of flatworms, the *Acoels* and the *Nemertodermatids* (Figure 6.1), very similar to each other, that were originally considered platyhelminths, but based on recent molecular evidence they both seem to be a separate, more ancient group. This means that they seem to be closer to the original "bilaterian animals" (more on that soon). However, this claim is still under intensive study and is somewhat controversial (but not that much).[10,11] I am sure that this is not the last word on flatworm classification. However, even though the Acoels and the Nemertodermatids are not "officially" platyhelminths based on current taxonomical rules, they are flatworms nonetheless, and their study will undoubtedly contribute to elucidating aspects of the evolutionary history of animals.

Regardless of how we choose to catalog flatworms, there are many aspects of these little guys that make them very intriguing members of the animal kingdom. To begin with, flatworms are thought to be among the most primitive organisms that display *bilateral symmetry*. What is this? This simply means that if you divide a physical object exactly in half from a certain plane or perspective, the right and left portion are mirror images of each other (Figure 6.3). Also, in addition to their left–right symmetry, they display distinct front and rear ends, as well as dorsal (back) and ventral (belly) regions. Organisms that show these traits are collectively called *bilaterians*. We humans happen to be bilaterians, as well as some 99% of all other animal species.

Bilateral symmetry is not the only game in town though. Primitive animals like sponges do not have any kind of symmetry to speak of, and despite that, even if they do not do much, they do survive very well, thank you very much (Figure 6.3). Many other types of organisms like cnidarians (jellyfish, sea anemones, and the like), for example, display yet an additional kind of symmetry, *radial symmetry*, where the organism displays no evident laterality (left or right sides), but they are symmetric around a central axis (Figure 6.3). Many types of animals

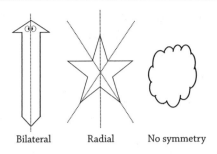

FIGURE 6.3. Representative animal symmetries.
Illustration by O. R. Pagán.

that display radial symmetry like starfish and jellyfish are also rather successful as well.

I am sure that you already know that in biology things are rarely simple; curiously, there are examples of organisms that can display more than one type of symmetry depending on their stage of development. For example, the echinoderms (starfish and associates) display radial symmetry in their adult forms, as we said before, yet their larvae display bilateral symmetry. Furthermore, animals displaying external radial symmetry, like cnidarians (jellyfish, hydra, and friends), can also present bilaterality in their internal structure. This just confirms that biology is a science of exceptions; life is much more complicated than we think. There are many "happy mediums." Consider humans, for example. Externally we are bilaterally symmetrical; however, internally we are not "complete" bilaterians, as shown by the example of the heart. We only have one heart, which is normally slightly "tilted" toward the left side of the body.

Let's explore bilateral symmetry some more. It seems that this kind of symmetry is an ancient characteristic. In evolutionary terms, conventional thinking suggested that the bilaterians display higher molecular and anatomical (morphological) complexity than radially symmetrical organisms. This assumption was based on two misconceptions.[12] The first one is that genetic complexity parallels morphological complexity. This tends to be true, but it is not absolute; it does not always apply. Organisms that arguably display higher physiological or morphological complexity do not necessarily have higher genetic complexity. For example, we humans have about 20,000 or so genes, about the same amount as the little *Caenorhabditis elegans* worm, which we talked about before. Very few people will argue against humans being more complex than these nematodes. This is just one example that shows that the number of genes in a species does not correlate with complexity; there are many more.

The second misconception is that genes, and by extension proteins, mutate at constant, predictable rates. This argues against consistent rates of change with

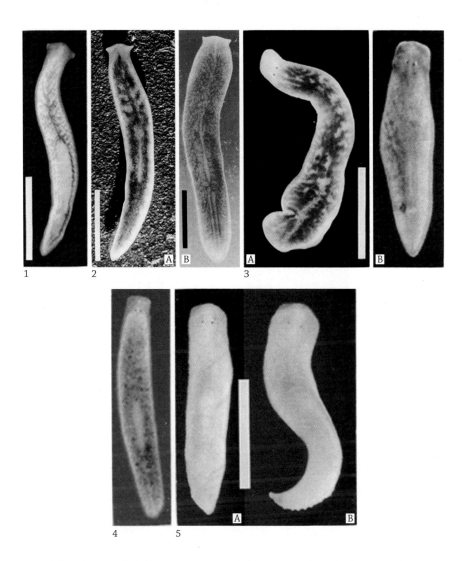

PLATE I. Color photographs of live specimens: Scale bars = 5mm. (1) *Polycelis sapporo* (loc. Chitose City, Ishikari, Hokkaidô: photo by Mr. Miyazaki). (2) *Seidlia auriculata*. (A: loc. Mt. Yamiozo, Ibaraki Pref., Honshû: photo by Dr. Chinone. B: a specimen fed by a fresh chicken spleen; loc. Nîgata Pref., Honshû: photo by Murayama). (3) *Phagocata kawakatsui*. (A: loc. Bottom of Lake Biwa-ko, Shiga Pref., Honshû: photo by Dr. Nishino. B: type loc. Ground of Kawakatsu's country residence in Kameoka City, Kyôto Pref., Honshû: photo by Mr. Miyazaki). (4) *Phagocata suginoi* (type loc. Soji in Kashiwazaki City, Nîgata Pref., Honshû: photo by Murayama). (5) *Phagocata papillifera*. (A and B: loc. A driven well in Mitsukaidô City, Ibaraki Pref., Honshû: photos by Mr. Miyazaki).

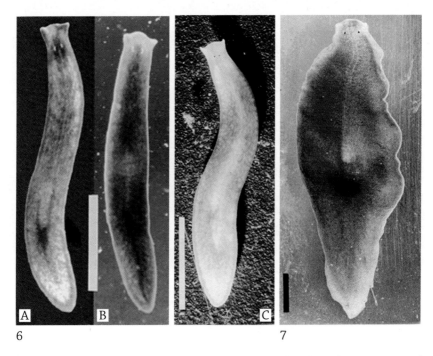

PLATE I. (CONT.) (6) *Phagocata vivida*. (A and B: loc. Nîgata Pref., Honshû: photos by Murayama. C: loc. Mt. Yamizo, Ibaraki Pref, Honshû: photo by Dr. Chinone). (7) *Bdellocephala annandalei* (loc. Bottom of Lake Biwa-ko: photo by Murayama).
Courtesy of Dr. Masaharu Kawakatsu. Used with permission.

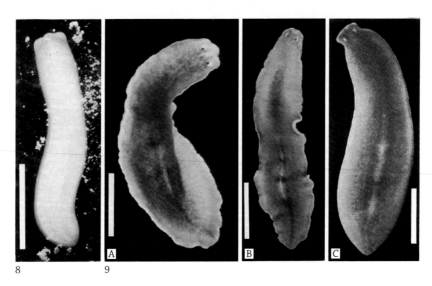

PLATE II. Color photographs of live specimens: Scale bars = 5mm. (8) *Phagocata papillifera* (loc. Mitsukaidô City, Ibaraki Pref., Honshû: photo by Dr.Chinone). (9) *Bdellocephala brunnea*. (A and B: loc. Downstream of the Amagase Dam, the Uji-gawa River (approximately 18km downstream from the outlet of the Lake Biwa-ko), Uji City, Kyôto Pref., Honshû: photos by Dr. Nishino. C: loc. Tsunan-machi, Naka-uonuma-Gun, Nîgata Pref., Honshû: photo by Murayama).

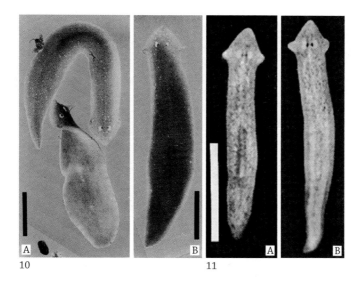

PLATE II (CONT.) (10) *Girardia dorotocephala* (loc. The Asakawa River, a tributary of the Tama-gawa River, Hino City, Tôkyô-to, Honshû: photos by Mr. Tsuruda).
Courtesy of Dr. Masaharu Kawakatsu. Used with permission.

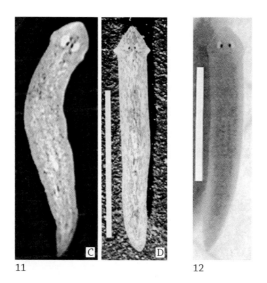

PLATE III Color photographs of live specimens: Scale bars = 5mm. (11) *Girardia tigrina*. (A: loc. Aquarium in Yokohama City, Honshû. B: loc. Aquarium in Nagoya City, Honshû: photos by Dr. Tamura. C: loc. Downstream of the Urakami-gawa River, Nagasaki City, Kyûshû: photo by Mr. Yamamoto. D: loc. unknown: photo by Dr. Kawakatsu). (12) *Dugesia austroasiatica* (loc. Aquarium in Tôkyô-to, Honshû: photo by Mr. G.-Y. Sasaki).

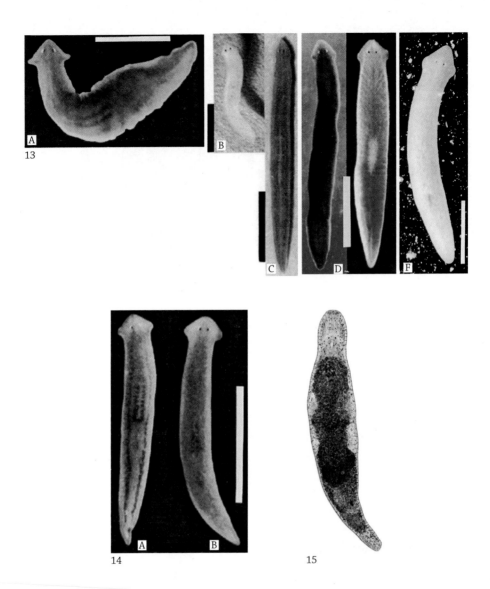

PLATE III (CONT.) Color photographs of live specimens: Scale bars = 5mm. (13) *Dugesia japonica*. (A: loc. Upper stream of the Urakanmi-gawa River, Nagasaki City, Kyûshû: photo by Mr. Yamamoto. B: loc. Bisai City, Aichi Pref., Honshû: photo by Mr. Hotta. C: loc. Izuhara City, Tsushima Islands, Nagasaki Pref., Kyûshû: photo by Mr. Yamamoto. D: loc. Tsunan-machi, Naka-uonuma-Gun, Nîgata Pref., Honshû: photo by Murayama. E: loc. Tamano City, Okayama Pref., Honshû: photo by Mr. Miyazaki. D: loc. unknown: photo by Dr. Kawakatsu). (14) *Dugesia ryukyuensis*. (A and B: loc. Downstream of the Urakami-gawa River, Nagasaki City, Kyûshû: photos by Mr. Yamamoto). Courtesy of Dr. Masaharu Kawakatsu. Used with permission. (15) The marine microturbellarian *Macrostomum lignano* (see text). Courtesy of Dr. Bernhard Egger, Institute of Zoology, University of Innsbruck, Austria. Used with permission.

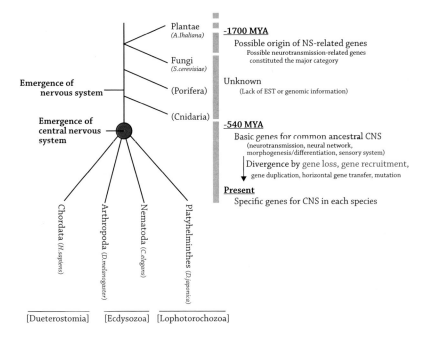

PLATE IV. A possible scheme of the process of evolution of central nervous systems. From Mineta K et al. (2003) Origin and evolutionary process of the CNS elucidated by comparative genomics analysis of planarian ESTs.
Proc Natl Acad Sci U S A. 100(13):7666–71. Copyright © (2003) National Academy of Sciences, U.S.A. Used with permission.

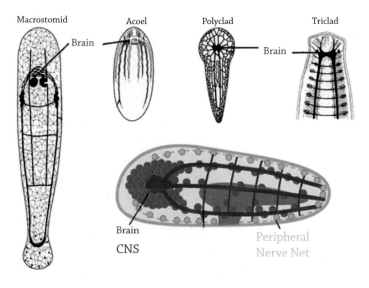

PLATE V. Examples of the specific arrangements of the central nervous system in various types of flatworms. From Bailly X et al. (2013) The urbilaterian brain revisited: novel insights into old questions from new flatworm clades.
Dev Genes Evol. 223(3):149–57. Copyright © 2012, Springer-Verlag. Used with permission.

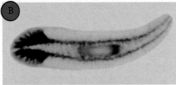

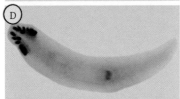
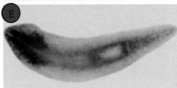

PLATE VI. The central nervous system of *Dugesia japonica* visualized using various types of selected molecular markers. From: Mineta K et al. (2003) Origin and evolutionary process of the CNS elucidated by comparative genomics analysis of planarian ESTs.

Proc Natl Acad Sci U S A. 100(13):7666–71. Copyright © (2003) National Academy of Sciences, U.S.A. Used with permission.

PLATE VII. Comparison of the central nervous system of three different freshwater planarian species based on the localization of a specific synaptic protein. Key: cg, cephalic ganglia (the planarian brain); vnc, ventral nerve cord; ph, pharynx (see chapter 6); tc, transverse commissure (the links between the nerve cords); lp, lateral process. Scale bar, 1 mm for *S. mediterranea*, 0.8 mm for *P. ullala,* and 0.5 mm for *P. felina*. From Cebrià F (2007) Regenerating the central nervous system: how easy for planarians!

Dev Genes Evol. 217(11–12):733–48. Copyright © 2007, Springer-Verlag. Used with permission.

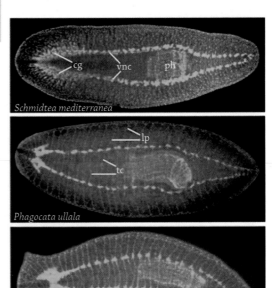

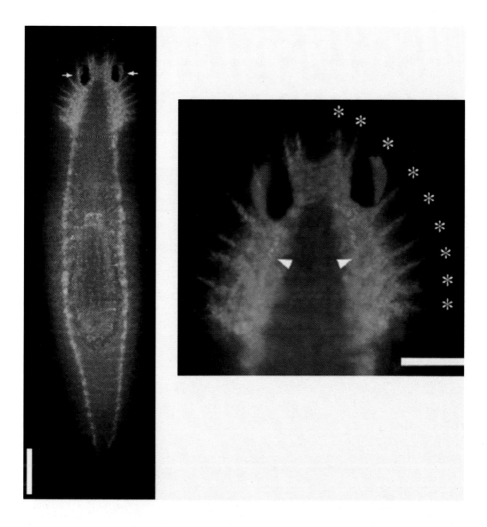

PLATE VIII. The brain and central nervous system general structure of the planarian *Dugesia japonica*. At left, the planarian central nervous system visualized by the fluorescence microscopy localization of specific proteins common in synapses and muscle tissue. Scale bar = 1 mm. At right, detail of the planarian brain, showing its nine branches. Scale bar = 0.5 mm. Modified from Umesono Y, Agata K (2009) Evolution and regeneration of the planarian central nervous system.

Dev Growth Differ. 51(3):185–95. Copyright © 2009 The Authors. Journal compilation © 2009 Japanese Society of Developmental Biologists. Used with permission.

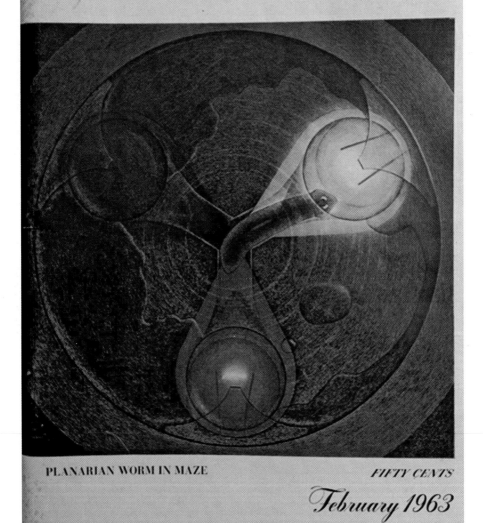

PLATE IX. The cover of the February 1963 issue of *Scientific American*, showing a planarian for the first and only time so far.
Copyright © 1963 Scientific American, Inc. All rights reserved. Reproduced with permission.

respect to molecular clocks. Again, this may be true in some cases, but not in others. Mutations are essentially random events triggered by environmental factors like radiation or chemical insults, as well as by mistakes in the gene replication process. These mutations are then nonrandomly selected for or against by natural selection (see chapter 1).

At any rate, recent work suggests that the Acoels and the Nemertodermatids are living representatives of the earliest bilaterians.[13] If this is correct, bilaterians in general seem to have been around for about 700 million years or so, although some molecular clock estimates (however, see earlier) suggest that bilaterians may date from about 1.2 billion years ago.[14,15] It seems that the earliest example of a bilaterian in the fossil record is of a rather complex organism, a mollusklike animal named *Kimberella*, which looked kind of like a flattened snail.[16] Of course, it stands to reason that there were other bilaterians simpler than *Kimberella* that simply have not yet been found in the fossil record. As we will see in the next section, invertebrate fossils are relatively uncommon; we are actually lucky that we even have invertebrate fossils available for study.

A predominant trait of the bilaterians that will prove to be especially relevant as we go along is that they tend to display *cephalization*, which is a reflection of the bilaterians' well-defined localization of front and rear ends. Cephalization is essentially the formation of a head including its associated sensory organs. We will explore this in further detail in chapter 9.

In evolutionary terms, the current thinking is that all bilaterians are descendants of a common "urbilaterian" ancestor. The term *urbilaterian* indicates the earliest common ancestor of all animals that display bilateral symmetry; in essence, it represents the last common ancestor of vertebrates and insects. The general consensus is that the Acoels mentioned before are the type of organism most similar to the ancestral urbilaterians, but again, the jury is still deliberating on this.

FLATWORM FOSSIL RECORDS

In general, invertebrates tend not to fossilize well, as they lack hard parts like bones, which are hardier and therefore more resistant to decay. However, many invertebrates, like insects, crustaceans, and so forth, possess an exoskeleton that, while not as hard as bone, has some resistance to decay, allowing more time and opportunity for fossilization to occur. Also, many invertebrates have hard shells, like snails and clams, for example, which fossilize rather easily. In the case of flatworms, we have none of these advantages. While it is not impossible that a soft organism can end as a fossil (in fact, we have quite a few examples of those), the probability of fossilization is lower due to the factors mentioned previously. There are other ways to create a fossil though; for example, an animal can leave tracks in soft terrain that, when hardened due to geological events, leave a stable record. Also, the outlines of whole organisms can be preserved in a similar way. Another type of very beautiful and well-preserved fossils are those conserved in amber, which is essentially hardened tree resin. A significant number of fossilized invertebrates in amber have been discovered and described.

As far as flatworms are concerned, fossil records are rather scarce. In addition to their soft bodies, most flatworms die in an unambiguous and somewhat dramatic way. Flatworms tend to *autolyze*, meaning that they essentially dissolve head first, like when evil guys die in a bad science fiction movie. The oldest undisputed direct fossil record of a turbellarian is an example preserved in amber, dated as coming from the Eocene period, meaning it was about 40 million years old (Figure 6.4A).[17] In 1975 a paper was published that reported a flatwormlike fossil in siliceous shale, apparently dated as Precambrian, therefore at least 540 million years old; its assigned scientific name was *Brabbinthes churkini*. Truth be told, it really looks like the outline of a flatworm (Figure 6.4B).[18] However, a more recent alternate interpretation of this fossil proposes that the fossil represents instead an example of an ancient sponge's needle or *spiculus*.[19] Some examples of planarianlike fossils were reported in rocks coming from the Miocene period (up to 23 million years ago).[20] Other examples of flatworm fossils include *Monogenea*-like fossil hooks (haptors; Figure 6.1) from the Devonian period (the Age of Fish, roughly 400 million years ago) and traces interpreted as flatworm tracks from the Permian period, close to 300 million years ago. More recently, fossil tapeworm eggs were found in a sample of shark coprolite dated from 270 million years ago (coprolites are essentially fossil feces).[21]

Flatworms are an ancient group indeed; they have been around for a long time!

WHAT IS A PLANARIAN?

As common as it is, the term *planarian* is rather nonspecific. It essentially means "flatworm," as in "look at that worm that happens to be flat." Therefore, at face value, the term is not very descriptive. This has prompted the suggestion of getting rid of the term *planarian* altogether (gasp!).[22]

I do not think that the term *planarian* will disappear anytime soon, as least in the vernacular. Remember that for most people *planaria, planarian,* or *planarians* refer to free-living, freshwater turbellarian, triclad flatworms of relatively small

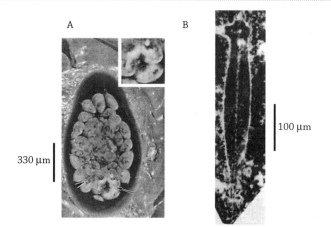

FIGURE 6.4. A. Flatworm fossil in amber (see text). The inset represents a detail of the worm's pharynx. B. Flatwormlike fossil found in siliceous shale, *Brabbinthes churkini* (see text).

A from Poinar G (2003) A rhabdocoel turbellarian (*Platyhelminthes, Typhloplanoida*) in Baltic amber with a review of fossil and sub-fossil platyhelminths. Invertebrate Biol. 122(4):308–12. Copyright, John Wiley and Sons. Used with permission. B from Allison CW (1975) Primitive fossil flatworm from Alaska: new evidence bearing on ancestry of the Metazoa. Geology. 3(11):649–52. Copyright, Geological Society of America. Used with permission.

size, usually not much more than one inch long, but often much smaller than that. The words *planaria, planarian,* and *planarians* are used more or less interchangeably; therefore, I'll use all three terms within the context of this book. Figure 6.5 shows a classification scheme of a planarian as a representative example of taxonomical hierarchy.

The first published instance of the word *planaria* that I was able to find appeared in the works of O. F. Müller (1776) in a book titled *Prodromus*, this word followed by a more elaborated description of the contents as part of the title, as was the style of the time (Figure 6.6A). *Prodromus* roughly means "sneak preview." In his *Prodromus*, Müller included a detailed listing and classification scheme of many types of invertebrates (Figure 6.6B). Please note that planarians were included within the mollusks, a group that now includes snails, slugs, squid, and octopi, among others. Again, taxonomy is very, very fluid.

There are hundreds of described planarian species; therefore, there is no specific description that applies to all planarians, and therefore the specific anatomical details of different planarian species will not be discussed here. There are plenty of general invertebrate zoology books that will do a much better job at it. However, whenever we think of planarians, usually the first thing that comes to mind is a distinctly shaped, more or less arrow tip–shaped head, although in fact, planarian morphology has a wide range of variability. They display various shapes, sizes and colors (Figure 6.7; Plates I, II, III). Also, they tend to exhibit an

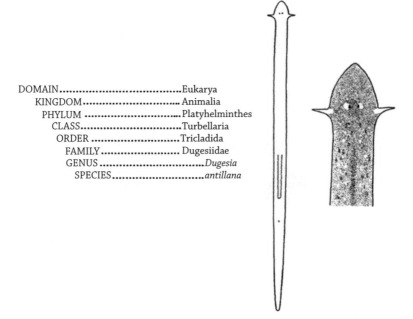

FIGURE 6.5. *At left*: Representative example of the taxonomical classification of a flatworm. At *right*: A flatworm from my native Puerto Rico, *Dugesia antillana*.

Adapted from Kenk R (1941) A fresh-water triclad from Puerto Rico, *Dugesia antillana*, new species. Occasional papers of the Museum of Zoology University of Michigan. 436:1–9. Copyright by the University of Michigan Museum of Zoology. Used with permission.

elongated, flat body and in many cases they have prominent "ears," misleadingly called *auricles*, which are not actually ears, as they do not detect sound. Rather, planarian auricles are actually more like noses, since they actually are chemoreceptive organs; planarians use their auricles to detect the presence of nourishing or noxious substances in their environment. In this way, they help the planaria decide whether to try to eat something or glide away from it. Most planarians also have simple eyes, called *ocelli*; in fact, they usually look cross-eyed or googly-eyed (Figure 6.7)! There are exceptions, of course; there are eyeless planarians, there are planarians with only one eye, and there are even planarians with hundreds of eyes. In all fairness, the term *eyes* as applied to planarians is a slight exaggeration. Their ocelli are best characterized as "eye spots" or "eye cups," as they lack lenses and other structures usually associated with proper eyes. That said, for simplicity's sake we'll refer to them here as "eyes." Because of their specific structure, planarian eyes are unable to form an image, but are nonetheless receptive to light intensity, which can provide the worms with a survival advantage. This is a very good example of how rudimentary eyes provide the baseline for the evolution of more complex visual structures.

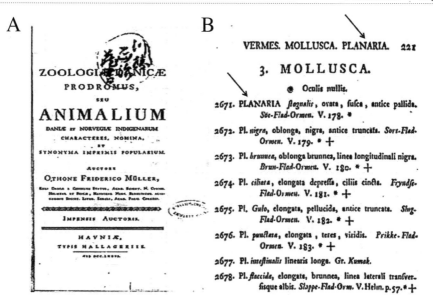

FIGURE 6.6. A. Othone Friderico Müller's *Prodromus* (1776). B. Detail of the page of the first recorded appearance of the term *planaria* in the scientific literature (indicated by the arrows). Pictures courtesy of Dr. Masaharu Kawakatsu. Used with permission.

Again, there is a wide variety of planarians out there, some species longer than others, some with more protruding auricles or with no discernible auricles at all, and so forth. Figure 6.7 shows representative planarian general shapes.

Despite being relatively simple animals, planarians can display surprisingly varied types of behaviors that can be easily observed and quantified. This is very useful when doing research with them. One of the most prominent behaviors of planarians is that they do not like brightly illuminated places; they actually flee from light. This is known as *negative phototaxis*, and if we think about it, it makes sense from the perspective of the worm. Freshwater planarians are small, slow organisms; they are also soft, with no hard exoskeleton, and they lack toxins or other means of defense. Basically the only option that they have is to hide, as the next best thing is to crawl away very, very slowly. Interestingly, many species do not seem to absolutely need their eyes to react to light. Decapitated planarians are still able to display negative phototaxis, albeit with lesser sensitivity as compared to intact worms. Even eyeless planarians seem to display some light sensitivity, as they can react to ultraviolet and visible light, but not to infrared wavelengths. It seems pretty evident that there are "extraocular" photoreceptors in these interesting organisms.

Planarians also respond negatively to touch. They change direction if gently poked with a toothpick, for example. They also seem to have a sense of direction

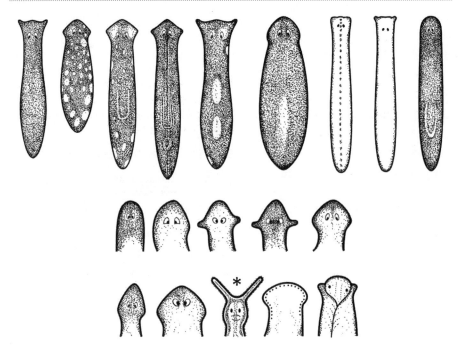

FIGURE 6.7. *Top*: Representative planarian morphology, not to scale. *Bottom*: Representative planarian head morphology. The worm marked with an asterisk is not exactly a planarian. It is technically an alloeocoel turbellarian; its scientific name is *Vorticeros praedatorium*, one of the few types of flatworms with tentacles.
Illustrations by A. G. Pagán.

and space orientation, which suggests that they are sensitive to gravity. For example, they display a so-called righting reflex; this means that if you flip them upside down while they are gliding on a surface or even at rest, they immediately "right" themselves. These little guys also display *rheotaxis*, which is essentially the capacity to react to a current of water. Specifically, planarians display positive rheotaxis; this means that they tend to move against the direction of the current. Further, they seem to be attracted to water currents. They appear to detect the water currents at the level of the head; they do not seem to react to currents aimed at their bodies.

These little guys are capable of other rather sophisticated behaviors. There is an account of a planarian attacked by a group of *Dilepti*, which are rather biggish unicellular protists (close to 1 mm long) that hunt by stinging prey with harpoon-like structures called cnidocysts, which contain a paralyzing toxic substance. The planarian attacked by the protists in this story reacted to the attack by writhing violently, as if trying to escape from them. After enduring the attacks for a while, in apparent "desperation"[23] it stuck its head out of the water, preventing the

Dilepti from stinging its head. Apparently the stings to its body did not bother the worm that much. The planarian kept its head out of the water until it dried off and died, but the rest of the body survived and was able to glide away, presumably able to regenerate another head and live another day.

This example illustrates another interesting planarian behavior: even though they are aquatic organisms, they will tolerate being out of the water for a while. They have been observed to get out of ponds and crawl around in the wet terrain nearby. I have personally observed examples of this planarian behavior when doing experiments with them. On occasion, when testing experimental compounds, some worms have been clearly not amused with whatever chemical or drug I expose them to; in some cases the worms have even tried to get out of the water. Several other interesting, more complex behaviors of planarians have been observed when the worms are exposed to drugs, particularly drugs abused by humans; these will be described in more detail in chapter 10.

As we saw earlier, planarians are carnivores, further, they are cannibals. Also, even though most species will happily feed on decaying matter, they can also be hunters as well. For example, they can prey on *Daphnia*, small crustaceans sometimes known as "water fleas." When planarians sense Daphnia, they secrete a sticky fluid that immobilizes the water flea and then the worm wraps itself around the flea just like a constrictor snake. Once in position, the worm extends its proboscis (also known as pharynx, Figure 6.2), a tubelike organ that contains its mouth, which doubles up as its anus (try not to talk about this on a first date...). In some planarian species the pharynx excretes digestive substances that seem to partially liquefy the insides of the prey and facilitate eating, in other species the pharynx "bites" and takes small chunks of tissue, and in other cases, if the prey is small enough, it swallows it whole. These little guys are very active hunters indeed; I have personally witnessed planaria attacking and feeding on *Triops*, another type of small ancient crustacean much bigger than planarians and possessing a hard exoskeleton. Planarians get under the Triops carapace, latch themselves onto the soft parts, and, well, there is no other way of saying it, suck it dry. In an ironic twist, Triops preys on the water fleas just mentioned, while planarians feed on both (a little more complicated than the circle of life). It has been hypothesized that planarianlike animals represent an example of some of the first hunters in the animal kingdom.[24] This hypothesis will prove relevant to our discussion of planarian nervous systems in chapter 9. It is widely thought that predation is an important source of evolutionary selective pressure. Before predators appeared, other factors such as disease and preadaptation to environmental conditions, among other factors, could have driven evolution. Predation is a well-known trigger of evolutionary arms races, and in general, predators need to plan their hunt, as opposed to just grazing, for example. Thus, it is logical that there will be a selective pressure for more "brain power" in predators compared to nonpredators.

EARLY WORK WITH PLANARIANS (1700S–1800S)

One of the earliest depictions of freshwater planarians was written by Peter Simon Pallas in his book *Spicilegia Zoologica*, published two years before Müller's *Prodromus* (Figure 6.8). In this book, he did not use the term *planaria*; he used the scientific name *Fasciola punctata* (Figure 6.8). However, his book has a figure with several drawings of worms; of these, three, maybe four of them clearly look planarianlike (Figure 6.8). The genus *Fasciola* is only used now to identify a type of parasitic flatworms.

In 1814 a lawyer named John Graham Dalyell published the book *Observations of Some Interesting Phenomena in Animal Physiology, Exhibited by Several Species of Planariae* (Figure 6.9A). This is a very readable book and is one of the classic works on planarians. It is thought to include the earliest scientific observations of planarian regeneration and is also thought to have the very first written description of a two-headed planarian. In this book, Dalyell wrote what is probably the most iconic phrase about planarians; he said that the planarian "…may almost be called immortal under the edge of the knife" (Figure 6.9B).

Dalyell's wide range of interests represent an excellent example of the attitude prevalent in the intellectuals of the time. He was a lawyer by training but had a strong interest in the natural sciences nonetheless. In addition to his interest, he had a meticulous attention to detail, which in my mind qualifies him as a scientist, no question about it.

Let me give you an example of Dalyell's attitude. On page six of the preface, referring to the research that he is about to describe, he writes: "Possibly, analogous enquiries have long ago been instituted by those more qualified for the task; and, what has appeared obscure in the nature of the animals now brought under consideration, may already be elsewhere satisfactorily explained." What he essentially said was that it was possible that others may have observed these aspects of planarians before him.

Spoken like a true scientist indeed. You just have to admire his humbleness—that's the way to do it! What I mean by this is that if you are completely honest, first and foremost with yourself, you can never be absolutely sure that you are the first person to describe a particular phenomenon in nature, and that is essentially the spirit of Dalyell's quote. The usual phrasing to express this feeling in more modern times goes something like: "To the best of our knowledge, this is the first observation of.…"

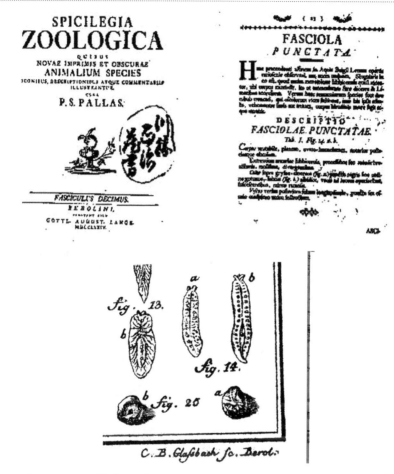

FIGURE 6.8. Peter Simon Pallas's *Spicilegia Zoologica* (1774).
Pictures courtesy of Dr. Masaharu Kawakatsu. Used with permission.

This is the absolutely best and most honest thing anyone can do when explaining or reporting science. I am sure that there are examples of this in other areas of science, but I found another instance of pristine scientific honesty by another planarian scientist (I know, I should say "scientist who works with planarians," but I could not help myself). This other example is from the biologist H. V. Brøndsted, who in 1969 published the book *Planarian Regeneration*, still a classic work on planarians. In the preface, Brøndsted recalls a comment that an unnamed friend of his told him once: "When you have done an experiment, then go through the (scientific) literature until you have found that your experiment have [sic] already been performed by others. Only then can you be sure that you have been through the entire literature".[25]

I simply must say it: planarian scientists are the best.

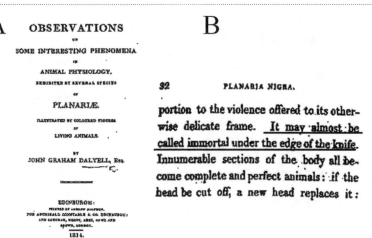

FIGURE 6.9. A. John Graham Dalyell's *Observations of Some Interesting Phenomena in Animal Physiology, Exhibited by Several Species of Planariae* (1814). B. Dalyell's famous phrase: "...immortal under the edge of the knife."
Pictures courtesy of Dr. Masaharu Kawakatsu. Used with permission.

Anyway, one of the observations that Dalyell records in his book is about the planarian proboscis (also called pharynx). Please recall that this is the organ through which planarians both eat and defecate. The pharynx reminded Dalyell of an elephant's trunk. In fact, taxonomically, elephants belong to the order *Proboscidea*; this term literally means "...the part located at front, which is used for eating." Modern elephants are the only extant (living) examples of this order. Other examples include extinct animals like mammoths, mastodons, and so forth.

Dalyell writes, "In imitation of the name bestowed on the trunk of the elephant, the extensile organ serving to imbibe the nutriment of many of the smaller animals, is called a proboscis...."

This seems to be the first time when the planarian pharynx was referred to as a proboscis. These connections are interesting. By reading Dalyell's book, one is able to learn the origin of one of the terms that we planarian researchers use on a daily basis.

Between 1822 and 1825, Dr. J. R. Johnson, a physician/naturalist, published two papers in the British journal *Philosophical Transactions of the Royal Society*, which incidentally, together with the French *Journal des Sçavans* (*The Scholar's Journal*), are widely considered the first scientific journals ever published. Both publications were first published in 1665 but strictly speaking, the Journal des Sçavans came first (January) while the *Transactions* was first published in March. The *Journal des Sçavans* ceased publication near the French Revolution. *Philosophical Transactions* is still being published and going strong and this is probably the reason why is considered the first scientific journal. It certainly is the longest running one.

Anyway, Johnson's first paper, "Observations on the Genus Planaria" (1822), in addition to reporting the planarian's cross-eyed morphology, described some of the basic behaviors that can be observed in the worms, namely, their locomotion and regeneration. He recorded the regeneration of both heads and tails, and he also observed the extension of their proboscis, although he did not associate them with the intake of food. At the time of this paper, Johnson was unaware of Dalyell's 1814 book. Johnson also reported the predatory nature of planarians by describing how they attacked an earthworm.

In his 1825 paper, *"Further Observations on Planariae,"* Johnson reported his observations of a double-headed planaria. He thought that his was the first observation of this, but he graciously stood corrected when he learned about Dalyell's book, and he gave Dalyell full credit for it, as well as for the realization that the pharynx was the planarian's feeding organ. In fact, he obtained a copy of Dalyell's book as a gift from the man himself (reportedly Dalyell's only copy). That was *gentlemanly*.

There is a very important point that I'd like to share. It is human nature to be attracted to novelty, to new things, and don't get me wrong, that is a good thing. That said, I very much disagree when people dismiss "old books" as if they were useless. I do understand this attitude a little better when multiple editions of books are available, but for many really important older books, like one hundred years old and even older, only one edition is available. This is true of books in any field, but this is particularly evident in the biological sciences. In many cases, these books include important information that is not available anywhere else. Many of those are very readable! Take Dalyell's book for example; it is very relaxed, entertaining reading, somewhat uncharacteristic of the science writing of the time. Who knows how many nuggets of valuable information are hidden in obscure books?

As much as I like old books, I acknowledge that science is about progress and the gain of new knowledge. In the next chapter, we will see some modern research that has taken advantage of the unique characteristics of planarians, but before we do that, I'd like to share three connections I have with three prominent "planariologists."

PLANARIOLOGISTS: THREE "PERSONAL" CONNECTIONS

As before, it is important to establish from the beginning what specific terms may mean. In this case, I want to define what I mean by "planariologists."

In previous chapters we have commented on the importance of model organisms in research. In these cases, we may have scientists who use a particular type of organism that is suited for a specific purpose because of a certain characteristic that proves useful (see chapter 2 for examples). However, the use of a specific kind of, say, animal to research a certain aspect of nature does not make the scientist a student of that animal. Let me explain myself a little better. We can use certain breeds of mice to investigate cancer, or wolves to gain knowledge on animal behavior. Similarly, many investigators have chosen planarians as animal models in regeneration biology, development, neurobiology, and pharmacology, among many other areas, but in general, their main objective is not to study planarians. I, for example, am not a planariologist but a pharmacologist who uses planarians as an animal model.

On the other hand, there are scientists who have made planarians and related critters the subject of their intellectual pursuits; it is precisely this type of scientist whom I will call "planariologists." This is, of course, an arbitrary definition that reflects my personal preferences; please note that this is not a formal scientific definition. That said, there are many distinguished examples of such scientists; therefore, a thorough account of them is way beyond the scope of this work. For that reason I have chosen to profile three of them with whom I have a connection of no more than one link. This means that I either have communicated directly with that specific planariologist or through a third person.

The first of these is Dr. Libbie Henrietta Hyman (1888–1969; Figure 6.10). Three sources of invaluable information on Dr. Hyman's life and works that I used in this section are included in the notes.[26,27,28]

Libbie Hyman is probably the most recognizable name in invertebrate zoology. She was born in Des Moines, Iowa, of immigrant Polish and German parents. She evidently did not have a happy childhood. In her autobiographical account, she described her early interest in nature yet was quite frank about her family life and how she was not given any encouragement to explore her intellectual interests, let alone any kind of family affection to speak of. The only person in her family for whom she had something like kind words was her father, and only to say that he "...was never cut out by nature to be a businessman. He was of a scholarly disposition...." In my view Dr. Hyman is a wonderful example of persistence, which is certainly an understatement, since she relates that her family actively dismissed her work and discouraged her from pursuing her academic interests. Worse, please remember that in the early 1900s the academic path for women in the sciences was "uphill" to say the least; that certainly did not help Hyman's aspirations. One of her teachers, Mary Crawford (a languages teacher mind you; never underestimate the power of an interested teacher), nurtured young Hyman's academic interests, eventually guiding her all the way to the University of Chicago, where she earned her BS in zoology in 1910. At the suggestion of Charles Manning Child (one of the preeminent planarian researchers of the twentieth century, whom we will meet in the next chapter), she enrolled in the University of Chicago's graduate school and earned her PhD under Child's supervision in 1915. In fact, she is arguably Child's most famous student, and guess what she worked with first? Hydra and *planarians*! The attitude of her family toward academia did not change even after Hyman obtained her doctorate, but that did not stop her from achieving her professional goals.

Dr. Hyman never thought of herself as the "research type," but one has to keep in consideration that either by temperament or indirectly from her family's attitude, she had a tendency to downplay her own accomplishments. Nonetheless, early in the course of her career using hydra and planarians, she noticed that these organisms were frequently misidentified in terms of taxonomy. That stimulated her interest in the area and she then became, in her own words, a "taxonomic specialist" of these organisms. As we will see, this was another of her characteristic understatements about herself; she was much more than a specialist, as she was a, if not *the*, world-class expert in these kinds of animals.

During her graduate studies she worked as a laboratory assistant in several zoology courses. She found the laboratory manuals lacking and therefore, after getting her PhD, she wrote her own manual titled *A Laboratory Manual for Elementary Zoology*, which was published in 1919 by the University of Chicago Press. It proved so useful and popular that she wrote a second, expanded edition in 1929 that also sold rather handsomely. Between these two editions, she wrote a more specialized manual, *A Laboratory Manual for Comparative Vertebrate Anatomy* (1922), which had a second edition in 1942 retitled *Comparative Vertebrate Anatomy*. Both were also bestsellers. She did all of this work while working as Child's assistant. Not bad for someone who, in her own words, "...never liked vertebrate anatomy...." I wonder, what could she have done with a subject she did like? This is kind of a trick question; we actually know what she did with the subjects that she liked.

Her academic love always was invertebrate zoology. Since she was receiving a rather sizable income from the royalties of her books and Child was retiring, she could afford to quit her job and dedicate herself to work on invertebrates. She worked (without pay) at the American Museum of Natural History in New York, where she was eventually made a research associate (still without pay, but with an office). Along the way, she produced a series of six books, more than 4,000 pages altogether, aptly titled *The Invertebrates*, written between 1940 and 1967. This series is a veritable tour de force in the subject of invertebrate zoology; each tome was dedicated to a particular invertebrate type. Without any assistance, she obtained, translated, and interpreted many primary papers on the subject and even took art lessons to prepare many of the illustrations herself. Along the way, she continued to publish original research and along the way inspired a couple of generations of students and established researchers interested in the invertebrates.

One of the researchers she helped and inspired was a scientist about forty years her junior, Professor Masaharu Kawakatsu, a self-described turbellariologist and naturalist (Figure 6.10). Dr. Kawakatsu is my "personal" connection to Dr. Hyman. I contacted Dr. Kawakatsu while writing this book to inquire about several scientific papers and I got much more than I had the right to expect. Dr. Kawakatsu honored me with his enthusiasm for this book and with his advice. Moreover, he provided me with invaluable hard-to-find papers and other materials as indicated throughout the text. Among the material he provided, he included a remembrance of Dr. Hyman that he wrote in collaboration with Prof. Taku Komai.[28] In his part of the *in memoriam* note, Dr. Kawakatsu spoke fondly of how Dr. Hyman treated him with the proper respect as a colleague and sent him many useful papers that further inspired him in his career. This is especially significant since he never met Dr. Hyman personally. He wrote: "Dr. Hyman taught me the true scientific spirit, namely that knowledge must be shared...." One of the tokens of gratitude from Dr. Kawakatsu to Dr. Hyman was the naming of a rare planarian species, *Dendrocoelopsis hymanae*, in Dr. Hyman's honor.[29]

Dr. Kawakatsu honored Dr. Hyman's role in his career in numerous ways. In the same way that Dr. Hyman helped and inspired him, he helped numerous students and researchers throughout his illustrious career. Among those he helped and inspired is another scientist about forty years his junior, the author of the book you are reading right now, for which I will be eternally grateful. In fact, many of the illustrations in this chapter were courtesy of Dr. Kawakatsu, as indicated in the figure legends.

Dr. Masaharu Kawakatsu was born in Asahi-mura Village, Kameoka-chô Town, Kyôto Prefecture, Honshû, Japan, on January 20, 1929. The earliest concrete record of his father's side of the family dates from about 400 years ago and indicates that he is a descendant of a Samurai. His interest in the natural world was initiated by his maternal grandfather, Dr. Ei'ichi Okajima, a medical doctor and an enthusiast of the natural sciences. Dr. Kawakatsu fondly remembers the conversations about nature and other academic matters he had with his grandfather, and he still has several of his grandfather's books, as well as the microscope Dr. Okajima used in his medical practice.

Young Kawakatsu embraced the study of nature very early in life. In fact, his first scientific paper was on some observations of plant life that he actually published in his school's journal at the age of thirteen![30]

He then attended Kyôto Normal College (now Kyôto Kyôiku University) as part of the Biology Study Group from 1946 to 1949. Kyôto Normal College, in collaboration with four other colleges in the area, organized a club for those interested in the biological sciences, the Biological Society of Kinki Normal Colleges. One of the most enthusiastic members of the society was Dr. Hisao Sigino, who worked on planarian regeneration. This was young Kawakatsu's first contact with a planarian researcher. Upon returning to Kyôto Kyôiku University as an advanced (graduate) student, Kawakatsu continued his research on planarians under the supervision of Dr. K. I. Okugawa, studying the regeneration of the asexual form of the planarian *Dugesia gonocephala*. This planarian species was the subject of Dr. Okugawa's doctoral dissertation.[31]

In turn, Dr. Kawakatsu earned his doctoral degree in 1961 with a thesis titled "On the Ecology and Distribution of Freshwater Planarians in the Japanese Islands, with Special References to Their Vertical Distribution." His thesis was subsequently published in 1965.[32]

By any standard, Dr. Kawakatsu's scientific productivity is quite significant. He lists close to 800 research papers, popular science articles, and educational publications overall, with approximately 1,100 coauthors from 16 countries. He has published about aquatic and terrestrial planarians in terms of taxonomy and cell and molecular biology. Upon his retirement in 1999, his vast specimen collection was deposited in the Zoological Museum of Amsterdam. This collection consists of approximately 22,000 glass slides and 1,400 preserved specimen vials.[33]

Finally, Dr. Kawakatsu was directly responsible for helping me find another connection with a distinguished planariologist, Dr. Roman Kenk (Figure 6.10).[34] Dr. Kenk was born in Ljubljana, Slovenia. Even though his family was not wealthy, he and his siblings earned advanced academic degrees and were very successful professionals. Dr. Kenk, in particular, earned a PhD in zoology at the Universität Graz, Austria, in 1921, when he was twenty-three years old. He worked at the University of Ljubljana for seventeen years. In the 1930s he met his future wife, a student from Puerto Rico named Ada Blanco y Pagán, at the University of Virginia in Charlottesville, where Ada was working toward a master's degree in botany and Dr. Kenk was an exchange professor on a Rockefeller Fellowship. After they married, they moved to his native Slovenia, where Dr. Kenk worked for a few years more at the University of Ljubljana until their move to Puerto Rico in 1938. He worked at the University of Puerto Rico (UPR) for about ten years and then moved to Washington, D.C., to work as a bibliographer at the Library of Congress. Upon his "retirement" in 1966 (when in good health, a true scientist never, ever retires), he was given a position as a research associate in the Department of Invertebrate Zoology at the National Museum of Natural History (Smithsonian Institution), where he was placed in charge of building the flatworm specimen collection and library, a position he held until 1987.

In an academic career that went from 1922 until 1987, Dr. Kenk published some seventy papers, most of them on planarians. It is worthwhile to note that many of those were detailed monographs on the subject and some of these were one hundred-plus pages long. Several animal species and at least two turbellarian genera (*Kenkia* and *Romankenkius*) were named after him by other researchers.

By all accounts, he was also a wonderful father and husband. Even so, ever the scientist, it was not an uncommon occurrence that family outings were at least partially planarian collecting expeditions. With his wife, Ada, and their daughter, Vida, they would set traps for planarians using the tried-and-true method of putting a piece of liver inside a glass container, poking holes in the lid, and leaving it in the water to be recovered the next day. According to Vida, her mother was far better than her father at trapping planarians.

He was also an affectionate, loving father. Frequently, when he was apart from his family, he would write letters to his little girl, but not just any kind of letters. He often composed them as a story, based on the life cycle of an organism. Of course, there was one about planarians. Here is an excerpt of it:

> Once upon a time there was a little soft slimy worm that lived in a cool creek not far from Richmond, Virginia.
>
> [Note from O. R. Pagán: at the time, Vida was living near Richmond, VA because of illness; she was about six years old.] Dr. Kenk continued:
>
> It was not quite alone, there were many worms of the same kind in the creek. They did not look like earthworms at all, they were not round, but flat as if somebody has sat on them. That's why those who knew them called them flatworms or planarians.
>
> Each planarian was, after all, a very peculiar creature. [Each] was dark brown all over and it had something like a head with two eyes on it and really looked cross-eyed. On the side of the head there were two little feelers that looked like ears. The planarian did not have a neck, nor legs. It did have a mouth, but that was put in the wrong place, in the middle of the tummy. After all, why should one have a mouth in the head if the food has to go in the tummy?

Dr. Libbie Hyman Dr. Roman Kenk Dr. Masaharu Kawakatsu
1888-1969   1898-1988   1929-

FIGURE 6.10. Three planariologists: Dr. Libbie Hyman, Dr. Roman Kenk, and Dr. Masaharu Kawakatsu.

Drs. Hyman's and Kawakatsu's pictures are courtesy of Dr. M. Kawakatsu. Dr. Kenk's picture is courtesy of Dr. Vida Kenk. Used with permission.

[Note from O. R. Pagán: Can you imagine that little girl's thoughts as she read that letter? After describing how they ate, Dr. Kenk narrated how a man cut a planarian's head off and how such head would grow back—better than a fairy tale, right?] Dr. Kenk concluded the letter as follows:

Now tell me if that is not fascinating. So small an animal and it can do more things than you would expect. A dog cannot grow a new head, not even a new tail, if you cut it off. And the little planarians can. And do you know who that man was who cut the planarians to pieces?
Your Daddy.

*...We will never understand the phenomena of
development and regeneration.*
—THOMAS HUNT MORGAN

7

PLANARIANS IN MODERN BIOLOGY

GENETICS

THE 1700S AND 1800S were really good times for science in general and for the life sciences in particular. In chapters 2 and 3 we briefly saw examples of this, like the formulation of the cell theory and the neuron theory. The 1800s also saw the origination of the theory of evolution by natural selection. These are only some examples of the explosive development of the study of the natural world, which gave birth to modern science. This era is sometimes known as the "Age of Wonder," and there are excellent books that chronicle these scientifically exciting times.[1,2]

One of the main questions regarding the evolutionary process as proposed by Darwin was how organisms were able to transmit their characteristics to their offspring, although nobody had the foggiest notion of how that happened. True enough, there were many ideas about the mechanism of inheritance with several degrees of logic, validity, and correctness. We are all familiar with Lamarck's infamous theory of the inheritance of acquired characteristics, which, sadly, is what Lamarck is best known for today. It is sad because he did so much more than that, including giving us the modern meaning of the term *biology* in 1812. Before Lamarck, *biology* meant solely the "study of man."

Another common mechanism proposed to explain the transmission of traits from one generation to the next was the concept of *blended inheritance*, which means exactly what it sounds like. This model proposes that the traits that came from each parent "blended," very much like mixing paint of different colors. Darwin himself spoke of the concept, but he did not originate it.

Both the theory of the inheritance of acquired characteristics and the theory of blended inheritance were incorrect (well, sort of; keep reading…). Lamarck's theory is mainly incorrect because traits acquired by an organism during its lifetime are not transmitted to its offspring. A rather commonly used example of this is that a person who trains as a bodybuilder will not have a child with particularly developed muscles because of such training. However, there is a recently described mechanism of inheritance named *epigenetics*, which is essentially inheritable changes in the DNA of an organism that are not related to changes in the gene's sequence. We can argue that this is indeed a kind of acquired inheritance, but this is a relatively new area of study, currently under research.[3]

With respect to blended inheritance, if it were correct, we would always expect an intermediate phenotype in the offspring compared to the parents' phenotype. For example, under this scenario, the offspring of a black wolf and a white wolf would always be gray. This could conceivably happen, not because of blended inheritance but rather as a consequence of the multiple factors that control heredity, many of them not well understood. Part of the difficulty when trying to understand inheritance is that, as we said before, many phenotypes are a consequence of a combination of the expression of multiple genes. In general, traits that can display continuous variability, like fur or eye color for example, are the result of something called *quantitative inheritance*. In other words, the range of phenotypes can go all the way from one extreme to the other. Height in humans, for example, is such a trait. People are not only five, six, or even seven feet tall; there is everything in between. However, biology is a science of exceptions, remember? There are actually examples in which something very much like blended inheritance indeed happens. For instance, the phenomenon of *incomplete dominance* occurs when neither of two given gene versions coding for the same trait (they are called alleles) dominates over the other. One specific (and very simplified) example is that of a common plant, the Four O'clock, which can express one gene coding for red flowers or its allele, which codes for white flowers. When both alleles are present, the flowers are not red or white, they are pink, an intermediate phenotype. A little closer to home, in human biology, the genetics of blood type also shows something like blended inheritance. In this case, it is called *codominance*. For example, if both alleles of a person code for "A," he or she will have type A blood. If both alleles are "B," the blood type will be B. So far so good, right? Now, if a person has one allele of each "A" and "B," there will not be an intermediate phenotype. Both alleles are expressed and therefore the blood type will be AB; no allele will dominate over the other. When none of the alleles are present, the blood type is O. This example is a little more complicated than the way it is presented here because of the intricacies of blood type compatibility and the fact that there are more subtypes in addition to the A, B, AB, O, the Rh factors, and so forth, but I believe that this is enough to illustrate the point.

It is important to point out that the aforementioned genetic scenarios usually occur in sexually reproducing organisms, in which one set of genes (half of the genome) comes from mom and the other half from dad. There are types of organisms where this paradigm simply does not apply, and even in sexually reproducing organisms there are often more than two alleles coding for a specific characteristic. As you see, there are many more layers of complexity in this area; as anything biology-related, genetics is a fascinating subject.

Historically, the scientific understanding of heredity also originated in the 1800s, even though no one in the scientific community at large seems to have realized it at the time. Genetics in its modern form started with the well-known story of Johann (Gregor) Mendel and his pea plants. Mendel's experiments formed the basis of the modern understanding of the laws of inheritance prior to the molecular biology revolution. Many of these principles are as valid today as they were about one hundred and fifty years ago. History records that Mendel published his findings in 1865 in a little-known journal (*Proceedings of the Brünn Society for the Study of Natural Science*). It is widely believed that at least partly because of this fact, his insights were not recognized until many years after they were published. Also, science historians generally think that several other reasons account for the lack of recognition of Mendel's work in his own time. Some of these reasons include the acute reluctance of Mendel to indulge in self-promotion (he was a monk, after all), the fact that Mendel essentially worked alone, and that his work was generally known as a speciation and hybridization work as opposed to work related to the more general phenomenon of inheritance.[4] Then, as now, terms and labels seem to matter. At any rate, Mendel's work was independently "rediscovered" in 1900 by three biologists, Erich von Tschermak, Hugo DeVries, and Carl Correns, who recognized it for what it was, and that moment in time likely initiated the modern genetic revolution, which is still ongoing.[5]

Once rediscovered, this better understanding of the general laws of heredity finally provided the beginnings of a mechanistic explanation of how organisms were able to transmit their genetic traits from generation to generation. This new understanding opened the door to the development of the so-called *new evolutionary synthesis* (*new synthesis* for short) or *neodarwinism*, where scientists combined the Darwin-style natural selection evolutionary thinking with Mendelian genetics. Eventually, the development of molecular biology added further details to the phenomenon of heredity, further contributing to this new synthesis. This is an ongoing effort; the nature of the relationship between genetics and evolution is still being refined as new aspects of it are discovered and reported.[6] I am sure that there is still a lot to be said on this topic.

FIRST THEY LIKED PLANARIANS AND THEN THEY DIDN'T

In the early twentieth century, the systematic search for the specific mechanisms of Mendelian inheritance was in full swing and several organisms were in a very real sense "auditioned" for the starring role in the story. One of them was our good friend the planarian! Prior to his pioneering studies in genetics using the fruit fly (*Drosophila melanogaster*), Thomas Hunt Morgan of Bryn Mawr College and later of Columbia University considered the planarian as a possible animal model to study genetics, development, and regeneration. By Morgan's own admission, his interest in doing planarian research was at least partly due to the influence of Dr. Harriet Randolph, of Bryn Mawr College, who published in 1897 the landmark paper "Observations and Experiments on Regeneration in Planarians."[7] Randolph's paper is important in this story since it masterfully summarized the state of the art of the planarian regeneration field at the time and described several experiments where Randolph systematically explored the regeneration of various amputation schemes in these organisms; along the way, her work got Morgan interested in planarians. Morgan took up planarian research with gusto; he wanted to use planarians to study genetics and regeneration. In fact, one of his first "scientific loves" was regeneration, not genetics, and he was exploring regeneration on a type of starfish at the time of his death. Anyway, during Morgan's most productive scientific years, his main competitor was Charles Manning Child of the University of Chicago. The relationship between Morgan and Child parallels in some ways the relationship and scientific rivalry between Santiago Ramón y Cajal and Camillo Golgi (see chapter 3). In the same way that Cajal and Golgi competitively explored the nervous system, Morgan and Child competitively explored the area of regeneration research. However, only one of them (Morgan) was awarded the Nobel Prize for his efforts (but not for his research on planarians but for his genetics research with *Drosophila*). An unintended consequence of this rivalry was likely one of the main reasons planarians were not the organisms of choice for the "new genetics" research. This was certainly a somewhat complex issue, with no single factor that seemed to be the sole cause of the gradual disappearance of planarian research in the early twentieth century, as *Drosophila* became the clear winner as the chosen model organism. A particularly thorough analysis of this episode in the history of science was published about twenty years ago.[8] Here I will only provide a very summarized overview of some of the main points in the Child–Morgan story as described in this article.

Probably the first point that needs to be made is that planarians were not in any way an intrinsically inferior animal model when compared with the fruit fly. Many of the advantages that are usually listed in favor of the fruit fly model also apply to planarians, namely, its portability and relatively easiness of culture. Actually, one may argue that planarians were more interesting because of their regenerative capacities, which *Drosophila* for the most part lacks. Ironically, it seems that the unusual regeneration capacities of planarians in a sense contributed to their "demise" as preferred animal models. Please recall that model organisms are generally chosen with a particular set of objectives in mind. In part I of this book we talked a little bit about the criteria by which scientists choose certain biological models over others. Some of these factors include how easy it is to work with them. This may sound trivial, but it is incredibly important; practicality is one of the main factors in scientific research. Other considerations that influence such choices are the specific characteristics of a given organism, like the large neurons in squids, whether an organism has well-defined genetics, or whether it reproduces fast, as well as many other factors.

In this sense, some researchers have said that the choice of model organism is heavily influenced by an organism's "exaggerated" traits,[9] consistent with the reasons mentioned earlier. However, I believe that it is extremely important to keep in perspective that these "exaggerations" are not biological anomalies, not by a long shot. They are only exaggerations from our limited, biologically provincial point of view. All biological traits have evolved over millions of years. They are kept by living organisms for no other reason than that they have helped their survival; upon survival, these organisms can reproduce and then the traits are passed on.

Another example of biological nomenclature based on human bias like the one we saw previously is when we talked about *extremophiles*,[10] which include a wide variety of organisms that thrive under environmental conditions that are "extreme" in terms of temperature and pressure, among other conditions. Again, they are considered extremophiles only because they live under conditions extreme *to us*. For them, though, they live in the perfect conditions to thrive. After all, again, they happened to evolve in such environments.

In terms of an ideal model organism, as far as the planarians and fruit flies story was concerned, there were two main scientific camps. The first camp, championed by Morgan, had as their main objective the elucidation of the principles of the transmission of genetic information—in other words, how the genotype was transmitted from generation to generation. From the perspective of this philosophy, the phenotype was only taken into account as the final product of the genotype; in a sense, it was almost incidental. On the other hand, the camp represented by Child was concerned with *how* the genotype's instructions created the phenotype; the main emphasis was on the mechanistic study of how we go from DNA to function—in other words, the study of *embryology*, or what we would call today *developmental biology*. The personal scientific outlook of Morgan and Child also played an integral role in this story. Both were good, even great scientists, but Morgan had a more practical bent; he emphasized data and results, without worrying too much about how those facts fit in the grand scheme of things. On the other hand, Child favored the integration of a wide body of experimental results within a general all-encompassing theory.

Due to the nature of the specific questions asked, it is fair to say that the scientific philosophies of Child and Morgan in essence needed different "dictionaries" to decode their particular inquiries about nature, in other words, different model organisms. In this sense, planaria was (and still is) an ideal subject to explore developmental and regeneration biology, while *Drosophila* was (and kind of still is) a useful subject to explore the fundamental mechanism of genetic transmission, although it needs to be recognized that in addition to its contributions to genetics, the study of the fruit fly has given us invaluable insights into fields like immunology, developmental biology, and, even more recently, behavior, among others.

With time, the research emphasis of the time shifted in favor of the exploration of the Mendelian theory of inheritance as opposed to the advance of developmental biology. Thus, since *Drosophila* was particularly useful in genetics, it slowly but steadily became *the* model. Planaria wasn't necessarily a bad model organism; the issue was that the unusual regenerative abilities of planarians were not obviously explained by the genetic thinking of the time. Truth be told, even today regeneration is not completely explained by modern genetics and molecular biology; this intriguing phenomenon is still being studied intensively.

This is not to say that Morgan was not interested in regeneration—far from it. Regeneration was kind of his first scientific love. He made important contributions to the field, including the distinction between two main regeneration pathways. He called the first one *morphallaxis*, which described regeneration as a process that involves not active cell division (proliferation) but the reorganization of existing cells, and the second one *epimorphosis*, which in turn explicitly requires cell proliferation. Let's not forget, though, that as I have said before in different ways, nature does not care about what we think of her or about our definitions of what she does. It is entirely possible that these two scenarios happen in a single animal. Moreover, there may even be cases when a combination of these two processes occurs; in other words, these two schemes (and maybe a couple of undiscovered ones) may not be mutually exclusive in an organism.

Anyway, what likely happened to Morgan as far as regeneration was concerned was that he got somewhat discouraged (and probably a little impatient, if you ask me) by the relative lack of significant results in the field; without molecular knowledge, the study of regeneration seemed to have reached a dead end. Therefore, Morgan was not very optimistic about the research progress on regeneration. He even went as far as categorically stating that "…we will never understand the phenomena of development and regeneration." This frame of mind, without a doubt, contributed to his final shift to *Drosophila* and genetics as his research focus.

As we have seen before, scientific research, as an activity done by people, is liable to be influenced by opinion, bias, or even whether you like or dislike someone. It shouldn't be like that but it happens; it is part of human nature. The professional differences between Morgan and Child extended to their personal relationship. In fact, their mutual personal dislike was a factor in the foundation of the journal *Physiological Zoology* by Child in 1928. Child's main motivation for creating the journal was that he felt that the *Journal of Experimental Zoology* (*JEZ*) rejected papers from his laboratory just because they were from his research group. Morgan happened to be on the *JEZ*'s editorial board and therefore Child believed (and was not shy to state it) that Morgan was at least partially responsible for the rejection of his papers. This is yet another example of scientific gossip at its best.

A particularly sad factor that has been blamed for the relative disappearance of planarians as animal models is the fact that many of Child's students were women; please remember that we are talking about a time when women were kind of systematically left out of prestigious academic (and any other for that matter) positions. However, note that by far, Child's most famous and accomplished student was Libbie Hyman, whom we met in the previous chapter. That said, she never held an actual academic position, which kind of proves the point. At any rate, because of these factors, not many laboratory heads eventually used planarians as their main animal model. In contrast, Morgan's students (men and women) are a true "who's who" list of the history of genetics in the twentieth century; most, if not all, of them worked with fruit flies. The end result was that in time, the predominant trend in research gradually shifted toward genetics research and therefore to the fruit fly as a model, at least in the United States. The story was different in Europe and Asia, where the influence of Child and planarians remained strong; therefore, the relative decrease in the use of planarians in scientific research did not quite happen in Europe and Asia, where the field remained popular.

SAGES OF REGENERATION

Many types of organisms possess some degree of regenerative capacities. This ranges from the healing of small cuts from injuries to the regeneration of whole body parts, both nonessential, like tails, and extremely essential, like nervous systems. There are even multicellular organisms (multicellular organisms are formally called *metazoans*) that can rebuild their own bodies when their cells are mechanically separated. For example, several species of sponges not only can survive the separation of their cells by a fine mesh, but also are able to reorganize themselves after dissociation, eventually regenerating a whole organism.

Some animal defense strategies take advantage of regeneration capacities. A common survival strategy that certain animals use to prevent predation is called *autotomy*, which is the spontaneous breaking of a nonessential body part that keeps a predator occupied while the intended prey escapes.[11,12] Autotomy is often but not always followed by regeneration. This practice is not limited to relatively small organisms like insects and earthworms. Animals like octopi are able to let go of one of their eight arms to avoid being eaten,[13] and even vertebrates use this survival strategy. I've seen this firsthand. I was born and raised in the tropical island of Puerto Rico, where small lizards of the *Anolis* genus are very common. We called them *lagartijos* (small lizards), and I witnessed their autotomy strategy many times in my childhood. Whenever a cat attacked them, for example, they frequently broke their tails off, which kept twitching; that frequently was enough to distract the cat and save the day for the little lizards. With time, the lizards grew their tails back—no harm, no foul. Lizard autotomy has been documented since the time of Aristotle, but I am pretty sure that the good philosopher did not observe this phenomenon in Puerto Rican *lagartijos*, just in Greek ones.

As far as we humans are concerned though, regeneration has a much more limited scope. With time, small cuts in our skin will heal, leaving a scar or even healing completely if the cut is small enough. Also, bones can regrow and fuse when broken and even severed fingertips can regrow in some cases. Healing after any type of injury is indeed a kind of regeneration. On the other hand, we cannot regenerate entire organs or appendages, with the exception of the liver (kind of). Sadly, when a person has brain or spinal cord damage due to accident or illness, the healing process is much more difficult or even impossible. You see, despite the ability of the brain to display plasticity (see chapter 4), the nervous system cells in vertebrates display very limited, if any at all, regeneration capacities.

Regeneration seems to be a trait widely distributed in the living world. Before we go any further, it is worth mentioning that plants are also very much capable of regeneration. You see, oftentimes we think of plants as hair, in the sense that we can get a haircut and the hair grows back. We do not usually think of that as regeneration. However, plants do regenerate indeed, and in fact, in the early days of biology, regeneration was primarily associated with plants. In an interesting twist, animal regeneration and its origins as an area of research was discovered in part by a curious naturalist trying to find out whether a certain type of organism called *hydra* was a plant or an animal by using regeneration as a metric. It was at this point in history when the formal and systematic study of animal regeneration in various organisms started, transitioning from mere anecdotal curiosities into an area of formal scientific inquiry.

In the 1740s Abraham Trembley, a Swiss natural philosopher, scientifically examined and documented regeneration for the first time. His story is just another example of the "Age of Wonder" era, where it was much easier to make an original fundamental contribution to human knowledge that did not necessarily have to be directly related to your own area of expertise. This does not mean that everything was new or even that no one thought or observed a particular thing; it simply means that that knowledge was lost or was not even recorded for posterity.

At any rate, Trembley did not have too much formal training in biology or arguably in any of the natural sciences to speak of. He was originally a mathematician, and at the time of his early biological studies his main job was as a private tutor for children of wealthy royalty. You would not think that this was the path of a first-rate naturalist, but Trembley was one indeed. Furthermore, it is widely recognized that science in its modern, recognizable sense started with Trembley. He argued that it was not enough just to observe nature; you must also experiment with it and further, record how you experimented with it so others could do the same and hopefully, within reason, replicate your results.

The organism that caught Trembley's curious eye, hydra, is the only freshwater member of the phylum *Cnidaria*, the same phylum that includes most jellyfish and corals. In a book like this one, mostly about planarians, it is not possible to completely understand how planarians contribute to our understanding of biology in general and of regeneration in particular if we do not at least take a look at the hydra story.

Essentially, hydra is little more than a slender tube that ends in a mouth surrounded by several tentacles. Trembley named it *hydra* in reference to the mythical Greek monster. In general, the main characteristic of cnidarians is that they are predators that capture their prey by injecting them with toxins through specialized cells called *nematocysts*. Once the prey is paralyzed, it is taken to the digestive cavity of the cnidarian using its tentacles. These tiny critters were described in the scientific literature for the first time by the very first microscopist, Anton van Leeuwenhoek, in 1702.[14]

Some varieties of hydra are bright green. We know now that this is due to a symbiotic association with green algae, but they did not know that in the 1700s; they just knew these hydras were green, and green is not a color usually associated with animals. Therefore, there was doubt as to whether hydras were plants or animals. In those days, the current paradigm was that for the most part, plants regenerated, while most animals did not. So on November 25, 1740, Trembley cut one of these guys precisely to try to determine whether they belonged to the plant kingdom or to the animal kingdom. His hypothesis was that if it regenerated, it was a plant; if not, it was an animal.[15] Lo and behold, it did regenerate! However, to his credit, Trembley did not immediately conclude that he was dealing with a plant; he was observant enough to notice that hydra displayed some behaviors that were not entirely consistent with what was known about plants at the state of knowledge at that time. For example, hydra wittingly captured prey with their tentacles, they were capable of reacting to tactile stimuli by contracting themselves, and they were also capable of active, purposeful locomotion. These traits suggested to Trembley that hydra had at least the rudiments of digestive and nervous systems, which were clearly animal characteristics. However, remember that we discussed aspects of plant behavior in chapter 3, including examples of carnivorous plants, which are pretty adept at actively capturing prey. It is unclear how much Trembley knew about carnivorous plants, but at any rate, he correctly concluded that hydra were animals, not plants.

An important aspect of Trembley's research illustrates the advantages of going beyond the artificial boundaries between biological disciplines. Even today, to some extent, botanists (scientists that study plant life) and zoologists (scientists that study animal life) rarely cross paths. Happily, Trembley clearly "crossed the aisle" between plants and animals, and because of that, biology in general and developmental biology in particular is much richer for it. Another important aspect of Trembley's history is that in contrast with the history of many other scientists and their discoveries, Trembley immediately realized that he had made an important discovery and worked very hard to make that known.

Upon further study, hydra proved to be true masters in the biological art of regeneration. They can regrow any lost body part. To illustrate how good these guys are at regeneration, it seems that a fragment composed of close to three hundred cells is capable of full regeneration.[16] Furthermore, even if a hydra is dissociated into individual cells—in other words, if all its cells are separated from each other through chemical or mechanical means—those cells will eventually reorganize into a fully formed hydra![17] This is reminiscent of the abilities of certain sponges as we discussed at the beginning of this chapter, with one significant difference: hydra possesses a fully functional nervous system—granted, a simple one, but a nervous system nonetheless. This is important, because as we have seen, nerves, at least in most vertebrate organisms, are notoriously known for their relative inability to regenerate after injury. As impressive as this feat is, it is important to understand that hydra has no obviously centralized nervous system. In other words, it has a nervous system, but no brain that we can recognize as such.

However, there is a type of animal that regenerates almost as well as hydra; moreover, this other organism also has a true brain. Even better, these guys are able to fully regenerate their brains upon injury or even after full decapitation.

Do I have to say that I am talking about planarians?

Many planarian species possess regeneration abilities to various degrees. In this section we'll mainly talk about the kind of planarians capable of full regeneration. In individuals of such planarian species, if you cut a planarian into two equal pieces, first, you will not kill it; second, each part will develop into a complete worm half the original size in about a week (give or take; again, depending on the species). Remarkably, the tail segment will regenerate a complete nervous system, including a fully functional brain! Not all planarians possess such power. There is a wide variability of regeneration capacities among planarian species. Some species are especially adept at it, some species regenerate but not very well or only under very specific circumstances, and yet some species do not regenerate at all. Incidentally, this is an important question from the perspective of evolutionary biology. You'd think that such a useful trait as regeneration would be more widespread in nature, but that is certainly not the case. This is puzzling; I am sure we all agree that the ability of regenerating missing body parts after injury is a rather useful capacity. However, even in organisms of the same class, as in the example of planarians, regeneration is not a universal trait. This is a very important question that is currently under intense investigation, in good part because it is a rather intriguing biological mystery. As far as developmental biologists can tell, there does not seem to be any general principle common to regeneration phenomena across the animal kingdom.[18]

At any rate, if you cut a planarian into 100 or even about 200 pieces (the "record" seems to be 279 pieces, but more on that soon), each piece has a fair chance of regenerating into a smaller yet complete organism, which can grow to full size if provided adequate nourishment.

A very important question that immediately comes to mind is the question of the minimal number of cells we need to fully regenerate into a whole individual; remember that for hydra, this number was about three hundred cells or so. However, if we think about it, in principle, a single cell would suffice. After all, most multicellular organisms that reproduce sexually originate from a single cell, the zygote (remember the birds and the bees?). The special thing about this original cell is that it is capable of giving rise to a whole variety of cell types, specialized to perform a particular function. That's one trick of nature we have not figured out yet. Incidentally, this is the main idea behind the whole science of stem cells.

Interestingly, in at least one case, it has been demonstrated that a type of parasitic flatworm can fully infect its host, regenerating complete individuals when the host organism is exposed to only one flatworm cell.[19] However, please note that in general a parasite is simpler than a free-living organism. This is not to say that parasites are less able to survive—quite the contrary! Parasites include some of the most successful organisms in nature; let's not forget that.

In contrast, free-living planarians, as far as we can tell, are unable to regenerate from just one isolated cell. So, from the point of view of planarian regeneration, one of the first evident and most interesting questions is, again, what is the smallest fragment of a planarian capable of full regeneration?

Please recall that from the late nineteenth century to the first few years of the twentieth century, Thomas Hunt Morgan, of *Drosophila* fame (we met him before), alongside his collaborators conducted a research program using planarians to study developmental biology and regeneration. One of Morgan's best-known experiments related to planarian research was his estimation of the smallest fragment of a planarian that was capable of regenerating a whole individual. He used an unusual yet rather creative approach. He essentially made a cardboard model of a planarian, drawn to scale. Then he proceeded to cut a piece of the worm while at the same time cutting the corresponding piece of the cardboard model. After that, he followed the worm's regeneration. He repeated the process until he determined the smallest piece that was capable of *some* regeneration. This fraction turned out to be 1/279. In contrast with almost every account of this experiment, this fraction was not capable of full regeneration, just partial. Oddly, he did not seem to explicitly report the absolute minimal fraction capable of completely regenerating.

Now here's the thing. Again, based on his experiments, Morgan determined that a fraction as small as 1/279 of the whole animal was capable of *some* regeneration, but by no means did a complete animal come out of it. Actually, Morgan never claimed full regeneration.[20,21] Nevertheless, this fraction is cited in practically every single paper or review article about planarian regeneration, and truth be told, it is the closest thing comparable to an article of faith among planarian researchers. Don't get me wrong: this fraction is quite significant. A 1/279 fraction represents a mere 0.36% of the total weight of the worm. By any standard, this is quite impressive indeed. For comparison purposes, it is like recovering a two hundred-pound human from about three-quarter pounds of flesh.

However, we need to keep in mind that Morgan's experiment is an estimation simply based on volume without taking into consideration how many cells were in that small fragment (in other words, cell density). Of course, Morgan estimated his figure using the cardboard model of the worm, which does not correlate very well with a cell-to-cell ratio. Also, it does not take into account the type of cells present in that minute fragment. Not all the cells in a multicellular organism are equal. Morgan did not seem to try to correlate this fraction with even an estimate of the number of cells present. Therefore, Morgan's fraction is a kind of overestimation of the true regeneration capacity of planarians—although it is impressive; don't get me wrong. Here's my reasoning.

The systematic estimation of the minimal number of cells required to allow for planarian regeneration had to wait until 1974, some seventy years after Morgan's experiments. In the appropriately titled paper "On the Minimal Size of a Planarian Capable of Regeneration," a research group reported that at least 10,000 cells were required for regeneration as defined by the appearance of ocelli.[22] In this paper, they estimated that the total number of cells in a typical planaria was about 200,000 cells. If these figures are truly typical, these 10,000 cells translate to about 5% of the total number of cells, an almost 14-fold difference from Morgan's ratio. A similar percentage (~5.7%) is obtained by dividing the minimal regeneration volume reported in the paper (0.08 mm^3) by an estimate of the total planarian volume (1.409 mm^3). So, Morgan's effort was a very good first approximation, but it was certainly not the whole story.[23]

The record for the smallest number of cells required for full regeneration in a nonparasitic flatworm seems to belong to a microturbellarian, the approximately 1mm long *Macrostomum lignano* (Plate III). It seems that these little guys are capable of regenerating from about 4,000 cells. However, let's keep in mind that these are smaller worms, composed of about 25,000 cells,[24] which means that *Macrostomum lignano* needs about 16% of its cells to fully regenerate. Still no mean feat though!

The story of *M. lignano* is a rather interesting one at several levels. This specific flatworm belongs to the Macrostomid taxonomic group, which includes close to one hundred species. These organisms are becoming increasingly popular in developmental biology because they are considered "basal" bilaterians, meaning closer to the phylogenetic origins of bilateral animals. Since they are microturbellarians, they are usually about 1 mm long, yet they have distinct organ systems of some complexity. Specifically, *M. lignano* was discovered as recently as 1995, in a culture of a related species, *Macrostomum pusillum*, used in developmental biology work.[25] With time, the *M. pusillum* was outcompeted by *M. lignano* in culture; it was later established that *M. lignano* was actually much easier to culture, and since then, many colonies have been established for research purposes.

Based on these facts, it is no surprise that flatworms are a preferred animal model in developmental biology and regeneration research. Can you imagine all the benefits we could get if we learned the secrets of regeneration, particularly of the nervous system? Think about the people with spinal cord injury or brain damage. Research on this subject has the potential to result in treatment strategies for these and maybe other related conditions. We will talk about nervous system regeneration and planarians in the next chapter, but in the meantime, let's talk about some aspects of what we know about how some planarian species are able to regenerate.

The history of planarian regeneration research is rich yet sadly not widely known. As we mentioned before, in contrast with the virtual elimination of the planarian model in the United States (as illustrated by the Morgan-Child story), planarian research remained strong and even popular in Europe and Asia. Between 1940 and 1960, a series of very dedicated and productive research groups in Europe and Japan took the proverbial planarian science flame and kept it alive. These groups emphasized the cellular aspects of regeneration, simply because the molecular tools were not available. The reawakening of the interest in planarian regeneration came about with the neoblast theory (neoblasts are rather interesting cells of which we'll talk more soon). As applied to planarians, the neoblast theory was articulated by Etienne Wolff and Françoise Dubois. This paradigm guided the philosophical agenda for planarian regeneration research over much of the 1940s to 1950s. They set the stage for much of the molecular aspects of planarian research. The best-known groups in this aspect hailed from Japan, Italy, France, and Spain. It is important to remark that many of these groups did work on the topic with relatively limited resources, which makes their achievements even more remarkable.

Multicellular organisms like you or me are composed of many, many cells of different types. Those cells cooperate to help us survive. What is remarkable, though, is that with the exception of the sex cells (sperm and eggs), all the other cells in a given organism have exactly the same genome. Now think about this: a cell taken from one of your eyes has exactly the same genes as a cell in your big toe, a liver cell, and so on—you get the idea. What is going on? The main reason for this is that during development and throughout the life of an organism (under normal circumstances), each type of tissue turns on some of its genes but not others. This selective activation leads to several things, including a phenomenon called *differentiation*. In this process, which starts with the zygote (our original cell, remember?), as cells divide, they become more specialized in terms of their function. This is a good thing, but there is a tradeoff. As cells specialize to perform a specific function, they tend to lose their ability to go "back to square one"; in other words, differentiated cells generally cannot "unlearn" what they know so they can "learn" to do something else. In some types of cells, they can even lose their ability to divide. This, of course, may be a problem when an organism is injured. Depending on the specific site and nature of the injury, if a cell type is so specialized that its dividing ability is compromised, recovery is compromised as well.

Some cell types retain their ability to "reprogram" themselves into making any other cell type. Cells like that are technically termed *pluripotent*.[26] This is a very useful trick as you may imagine. This is the kind of trick that any organism capable of some regenerative capacities has learned to do, at least to some extent. In vertebrates, including humans, this type of cell is called a *stem cell*. As we saw before, hydra is a true genius at regeneration. The pluripotent cells in hydra are called *interstitial cells*. In planarians, pluripotent cells are called *neoblasts*.

Neoblasts are really interesting cells, which have been known for a long time. They were described for the first time more than one hundred years ago in annelids (segmented worms) by Harriet Randolph, one of Morgan's associates who, in fact, started experimenting on planarians before Morgan. In planarians, neoblasts are the only cell population that undergoes cell division and are the responsible agents that control regeneration. Recently, a group based at the Massachusetts Institute of Technology (MIT), led by Dr. Peter Reddien, demonstrated that if you destroy the regeneration powers of a planarian using X-rays (radiation), it can recover the ability to regenerate if "seeded" with one—and only one—neoblast.[27]

Tough and resourceful little cells indeed.

OF PLANARIANS AND GENOMES

One of the most rewarding experiences for a person is to have a pleasant conversation with a like-minded individual. To us scientists, a conversation with a fellow science enthusiast can bring us to the level of junior high school students.

By this I mean a level of enthusiasm for the subject that sadly, many budding scientists lose as they grow up.

I recently had that kind of conversation with two such scientists, Dr. Alejandro Sánchez Alvarado of the Stower Institute in Kansas City, Missouri, and Dr. Phil Newmark of the University of Illinois at Urbana-Champaign. These two scientists are widely regarded as the two main protagonists on the study of planarian regeneration in terms of molecular biology. In my conversations with them, they kindly gave me some historical insights based on their personal perspectives and experiences on the use of planarians as organisms to explore developmental biology in general and the phenomenon of regeneration in particular. I will start with my conversation with Dr. Sánchez Alvarado. He is one of the top researchers in the field despite coming to planarians relatively recently; his first publication in the area was in 1998. He was also one of the leaders in the efforts to sequence the genome of a particular species of planarian, *Schmidtea mediterranea*, whom we will meet in a little while.

The main ongoing professional interest of Dr. Sánchez Alvarado is the biology of stem cells, particularly how they influence the phenomenon of regeneration. At roughly the same time when he began pondering the possibility of formally studying this in a scientific way, he also began to experience the type of influences that will cause a less driven scientist to stop, reassess, and change research directions (not unlike a planarian that encounters an obstacle while happily gliding underwater). For example, many people whom he asked for guidance about the field of regeneration told him that relatively few organisms regenerate, that regeneration was a mere biological curiosity as opposed to a widespread phenomenon and therefore that his time was better invested in alternate pursuits. Well, he would have none of that, so he simply did what a true scholar does, hit the books. What he found was a true mystery, not only was regeneration not as rare in nature as he was told, but in virtually every class of animal there is at least one representative capable of pretty decent regeneration. Intriguingly, within the same class of organisms, there were some that were able to regenerate while others did not.

He also found that the mainstream thinking pertaining to regeneration saw it as the "rewind and replay" (that's a 1980s phrase; look it up) of embryonic development, yet regeneration can be seen as an "adult animal" phenomenon. Moreover, by definition, since all adult animals have gone through embryonic development, regeneration would simply be, again, a matter of restarting development and that is simply not what happens.

Therefore, very early on Dr. Sánchez Alvarado realized that vertebrate animals were not optimal regeneration models for a couple of reasons, one of them being that relatively few types of vertebrates are capable of regeneration, and even when they did show some regeneration, it was never complete. In this sense, the people who advised him not to pursue this biological problem were kind of right, if only as far as vertebrates are concerned. The second reason for not using vertebrates to study this problem pertains to the pretty much underexplored molecular biology of the phenomenon itself.

At that point, he began to explore the possibility of finding an invertebrate model organism suitable to study regeneration. One of the main objectives he had in mind about this was the full sequencing of a representative genome of a regeneration-able organism to try to understand the phenomenon at the molecular level. He had four main criteria for the selection of an ideal animal model to study regeneration, a wish list so to speak. His ideal model would have:

- The capacity of full regeneration
- A diploid genome
- The capacity of undergoing sexual reproduction
- A relatively small genome

A fifth implicit criterion of his was that the animal must be basal enough, preferably to the bilaterians (meaning closer to the phylogenetic origins of animals, as we saw before). For a while he considered hydra, but developmentally and for technical reasons this organism is too simple. During the course of his literature review, he came across the writings of Harriet Randolph and Thomas Hunt Morgan, especially the ones pertaining to regeneration and planaria. He was understandably impressed and rightfully so, as many of the unanswered questions about regeneration that Morgan considered important are still valid today. He shared with me that he still vividly remembers the exact moment when he realized that planarians were it, that they were *the* model to study how some animals are capable of regenerating. He was a postdoctoral researcher at the time, and when he talked with his postdoctoral advisor about using planarians, his advisor emphatically tried to discourage Sánchez Alvarado from the idea by saying that planarians were almost impossible to manipulate and that frankly, they were rather "boring." I, for one, am glad that Sánchez Alvarado did not follow that advice. With time, he narrowed down his choice of model organism to one, the planarian *Schmidtea mediterranea*, an Old World species that pretty much matched Sánchez Alvarado's wish list. This species was first described in 1975 by the group of Dr. Mario Benazzi in Italy and was originally called *Dugesia lugubris* and even later *Dugesia mediterranea*.[28] In the close-knit community of *Schmidtea mediterranea* researchers, they are nicknamed "Smeds."

Chance has a prominent role in the sciences. There are many tales that give us an account of "happy accidents," which you may have heard defined as serendipity. What makes science different than blind luck is that alert scientists are able to seize the moment and take full advantage of it, not unlike the relationship between the random nature of mutations and the highly specific role of selection in evolution. Unbeknownst to Dr. Sánchez Alvarado, another like-minded scientist also had in his sights planarians, but not just any planarians, the Smeds, as an experimental organism. In our conversation, Dr. Sánchez Alvarado called this scientist, Dr. Phil Newmark, his "scientific brother."

Newmark met the Smeds first. He was a postdoctoral fellow of Dr. Jaume Baguñà, a seasoned planarian researcher at the University of Barcelona (Spain) who had been working on planarian regeneration since the early 1970s; in fact, Dr. Baguñà was a coauthor in the 1975 paper where *Schmidtea mediterranea* was first described (see note 28). The Department of Genetics at the University of Barcelona was then and still is a "hotspot" in planarian research. Newmark originally contacted Dr. Baguñà in 1989, while still a graduate student, because at the time, that was the only laboratory in the world that was beginning to study planaria at the molecular level. However, they were mainly interested in working on *Dugesia (Girardia) tigrina*, which did not seem to be amenable to molecular studies. Newmark suggested that his hosts use Smeds instead, basically for the same reasons that Sánchez Alvarado had, but at the time, it was not meant to be. In fact, Newmark is a coauthor of two 1996 works from Barcelona that used *Dugesia* instead.[29,30]

Newmark and Sánchez Alvarado met by chance in London, England, at the 1996 meeting of the British Society for Developmental Biology. A friendship soon ensued and they began to make plans to work together. The matter of using Smeds as the model to pursue their goals was preeminent in their thoughts. They both needed the worm, and they both wanted the worm, so in 1998 they went to Spain to get the worm.[31] In time, Sanchez Alvarado and Newmark, alongside Peter Reddien, whom we saw before, are credited with bringing planarian biology into the genomics age. As a consequence of their efforts, a representative genome of *S. mediterranea* is already available to researchers.[32,33] This will undoubtedly prove an invaluable resource to the rising number of investigators choosing planarians as their model organism in a variety of biological disciplines.

Two other species of flatworms will soon join the list of living beings whose representative genome has been sequenced, *Dugesia japonica* (first described by Dr. Masaharu Kawakatsu, who we met in the last chapter) and our friend the marine flatworm *Macrostomum lignano*.

It is important to point out that all of this did not happen in a vacuum. Remember that planarians remained a popular animal model in Europe and Asia, were *S. mediterranea* and *D. japonica* were extensively studied in regeneration by many research groups led by outstanding researchers.[34-37] These groups provided and are still providing a fundamental body of research that undoubtedly influenced Newmark, Sánchez Alvarado, and Reddien, as well as many others, in the finest tradition of scientific research. An especially good book that presents the early story of planarian regeneration research up to the late 1960s was written by another prominent figure on planarian research, Holger Valdemar Brøndsted (1893–1977; we saw him in chapter 6).[38]

As before, we just saw a minute fraction of all the exciting research done with planarians. Just to illustrate the point, one aspect we did not touch upon here pertains to senescence (aging) and how many planarian species seem to be in a state of "constitutive adults." That is, they grow and develop and then essentially stay young. I have a feeling that I do not have to argue too much to convince you that this is an extremely important area of research. Should you wish to know more about it, I refer you to several articles written by some up-and-coming researchers working directly in this area.[39-41] These are exciting times to be a planarian researcher!

"May you live in interesting times" is supposed to be an ancient Chinese curse, but for planarian researchers, this time in the history of science is pretty interesting, in a very good way. It promises an age of significant biomedical discoveries in general, with particularly interesting implications for one of the most exciting scientific disciplines, the neurosciences. These little worms are already proving themselves as important players in this area, as we will see in chapter 9. In the meantime, allow me to offer you an interlude that presents some aspects of the popular culture where we can find planarians. This is coming right up in chapter 8.

> *Behavioral Scientists (B.S.) have proved that flatworms (*Dugesia tigrina*) can do almost anything better than we (*Homo sapiens*) can.*
> —FREDERIC WAKEMAN[1]

8

PLANARIANS IN THE POPULAR CULTURE: THE ARTS, SCIENCE FICTION, FANTASY, AND HUMOR

PLANARIANS IN THE POPULAR CULTURE

AT THIS POINT in the book, it is pretty evident that I think planarians are quite interesting living beings. From my perspective as a scientist and university professor, I have worked with planarians for several years as they represent fascinating research subjects. As we will see in the next two chapters, it is remarkable that organisms with such simple nervous systems display a series of relatively complex behaviors and pharmacological responses similar to much "higher" organisms including ourselves. Thus, from this point of view, these little guys are an excellent model to try to understand how our own brain works.

The previous paragraph lists several scientific reasons why planarians are great laboratory subjects, but one does not have to be directly (or at all) involved in the sciences to be fascinated by planarians. These worms have captured the imaginations of people from widely different walks of life. In addition to catching the interest of science, they have inspired the graphic arts, poetry, and science fiction, of course. Moreover, their unique characteristics have served as literary tools to explore aspects of the human condition such as the ever-present pursuit of youth, as well as the (non?) inevitability of mortality. Also, it is not surprising that as with absolutely everything in life, in addition to the various forms of artistic expression described previously, planarians have been seen in a humorous light. There are many wonderful examples of this. I have chosen just a few of them as an interlude before we continue our scientific exploration of these fascinating beings.

PLANARIA

Run for a dictionary if you're the type that needs
a definition straight off.
But to learn *planaria* the way my brother did
you'll have to put on lug-soled boots
and carry two pounds of dead animal
into scrub oak. You'll have to hike for an hour
saying a prayer to the hog
that gave up his liver for science.

You'll have to hum something bluesy
to purge your blood of intention,
then whistle off-key to trick lost paths
into remembering your feet.
Planaria—taking two steps for each of his
and the backs of his knees opening
and closing like baby mouths. Yes, planaria—
kneeling to hook liver to a rusty lure.

And again planaria—sploshing that slippery
bundle of blood into a stream.
We could have waited, but sometimes it's better
to walk away from definitions. Better
to unroll sleeping bags, as if they were sleeping
ghosts, better to let smoking coals argue
things out with a tin foil dinner.
When we returned, moon gleamed everywhere.

My brother lifted the liver from the stream.
Its blackness looked like the blackness
of before, but dripping with water.
Water and planaria. There—I've said it.
"An order of related small soft-bodied free-living,
turbellarian flatworms moving

by means of cilia." See what I mean
about definitions? Like leeches, only smaller.

Planaria cannot see light.[2] They live on water
and blood. With mild shocks,
and with training, you can lead them through
an obstacle course in your bathtub.
If you cut them just right, they will grow
two heads. That's what my brother lowered
into a coffee can and snapped the lid over.
He carried that squirming song through the dark.

When I couldn't stay warm that night,
my brother zipped our sleeping bags together
and we fell asleep counting
constellations we could point to but not name.
All night, his fire smell kept oblivion
close. All night, that coffee can to the left
of our heads, a rock securing the lid,
a rock failing to hold the word *planaria*.[3]

PLANARIAN MAN

A big part of being a child in our culture is to enjoy comic books at one point or another in our lives. Moreover, you do not have to be a kid to enjoy this type of literature. The widespread popularity of comic books is undoubtedly one of the reasons that many comic books are adapted to movies. The allure of those movies comes in great part because people love seeing their favorite characters and their world come to life. One has to search very hard to find someone in our culture who has not heard about Superman, Batman, or the X-Men, among many other superheroes.

> Then again, there are some superheroes that are not as well known, like *Planarian Man* (Figure 8.1).

Yes, there is a superhero called *Planarian Man*. It is the brainchild of Neal Obermeyer, a very accomplished editorial cartoonist who hails from Omaha, Nebraska. His career started at the *Daily Nebraskan*, the newspaper of the University of Nebraska. His work has also appeared in the *Auburn Press-Tribune*, the *Omaha City Weekly*, the *Lincoln Journal-Star*, and the *San Diego Reader*. In 1996 he began drawing his series "The Adventures of Planarian Man," which was originally published in the *Auburn Press-Tribune*.

Planarian Man is a crime fighter, and he is, of course (what else?), half planarian and half man. He came to be when a young boy cut his finger when slicing a planarian as part of a high school biology laboratory experiment. Through a series of the kind of improbable events that oftentimes happen to people destined to become superheroes, a slice of the planarian came in contact with the cut in the boy's finger and became permanently attached to it. Over time, the boy began to change and turned into Planarian Man, the crime fighter.

The *Planarian Man* story touches upon some very interesting philosophical ideas. For example, in chapter 7 we talked about the planarian's ability to regenerate a new body from a small piece. In the course of his adventures, Planarian Man lost body parts while fighting crime, which was no biggie to him, as he was able to regenerate the missing pieces. As with real planarians though, his missing parts in turn regenerated into whole individuals. One of those individuals became Planarian Man's arch enemy, the Mad Dr. Planarian, who is an excellent metaphor of the evil that all of us are capable of. In this sense, Dr. Planarian is somewhat reminiscent of Robert Louis Stevenson's classic tale "The Strange Case of Dr. Jekyll and Mr. Hyde," as Dr. Planarian can also be seen as a representation of one's worst enemy. Think about it: who would be your worst enemy? An evil you, plain and simple, who in addition to being evil, knows everything there is to know about yourself in a rather intimate manner, strengths and weaknesses alike. This is a powerful concept that is extensively explored in the science fiction/comic book world.

The idea to create *Planarian Man* came to Neal while he was sitting in a high school biology class, listening to a lesson about planarians. What he found most remarkable was the story behind the 1960s memory experiments on planarians. We will very briefly see these memory experiments in chapter 9, but for now we can say that these were a series of studies where planarians were trained to react upon exposure to a certain stimulus. Furthermore, there was some evidence that at the time seemed to indicate that a planarian regenerated from a missing part seemed to retain, at least partially, the training of the original worm. It gets better; planarians are cannibals, so some of the memory experiments done in planarians involved training a batch of planarians, which were then chopped (yes, it is what it sounds like) and their pieces fed to untrained worms.

Anyway, a factor that was surely in the back of Neal's mind and that contributed to the creation of the character is that he lost the tip of one of his thumbs in a farm accident during a summer job; that surely put him in a frame of mind that when combined with what he learned about planarian regeneration gave life to *Planarian Man*.

You have to love these creative minds. You can find more about Neal and his work at the following websites:

http://nealo.com/
http://www.cheeksofgod.com/planarian-man/
http://www.planarianman.com/

PLANARIANS AND THE REIMAGINED *BATTLESTAR GALACTICA* TV SERIES[4]

You do not have to be a scientist or even like science in general to like science fiction (SF). That said, if you like science and are lucky enough to work in the area, it is very likely that you also like science fiction; it kind of comes with the territory. In other words, a love for science frequently means a love for SF. It also happens the other way around. Many scientists are originally drawn to science through science fiction. Part of the reason for this is that a true scientist has a well-developed sense of curiosity about nature, which usually means that a practicing scientist extends the frontiers of knowledge with each experiment. In this sense, it is very easy to see why many scientists identify with SF, as SF deals specifically with the current or imagined frontiers of knowledge. More often than not, these frontiers are seen in the light of the human experience.

Doing science is exciting. When a scientist gets data, it is entirely possible that the *only* person in the world who knows that little piece of information about the universe is at that given point in time the person who worked on it; trust me, it is a *great* feeling. Further, even if you are not the first person to know a certain fact, it is *new to you*, and that is a pretty good feeling in itself. Another meaningful thing about research is that with each new discovery, a whole series of new questions are immediately brought to light. As a consequence of this, explicitly or implicitly, the very first question that comes to the mind of the researcher after learning something new is "What if...?"

Therefore, a logical extension of this fascination with nature is to wonder about the "what ifs" of the universe in general. Usually scientists relate these "what ifs" of the universe with the future of science, with new discoveries, with

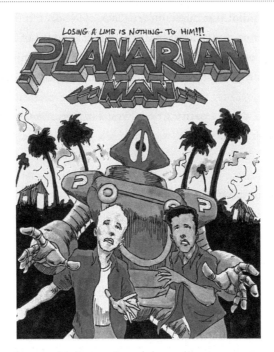

FIGURE 8.1. *Planarian Man* comic book example.
Courtesy of Neal Obermeyer. Reprinted with permission.

the development of new technologies, or even with new, unexpected ways of looking at nature itself. This curiosity can also be about novel, unforeseen implications, as well as about applications of current knowledge. So you see, we can consider science fiction as the literary exploration of the whole universe's "what ifs." Another way of looking at this is that science fiction is about the speculation of the kind of knowledge that humans cannot, for a variety of reasons, experience firsthand (yet).

My all-time favorite science fiction TV series (so far) is the new *Battlestar Galactica* (*BSG* from now on), a SF series that aired from 2004 to 2009. This series was a remake of a rather popular 1970s series of the same name. I absolutely love *Galactica*, but make no mistake, I am also a proud Trekker (or Trekkie, I am okay with both terms), and I loved *Babylon 5* as well as many other SF series and movies (is *the force* strong in you?). In fact, I prefer any science fiction TV show or movie—the good, the bad, and the just plain ugly—over any other genre, especially so-called reality TV, and I suspect that I am not alone.

The reimagined *BSG* was quality television; I do not think that many people will disagree with this statement. The new *BSG* had obvious entertaining features: good story lines, action, spaceships, the works. Moreover, for the most part its scientific background was sound.[5] About the "science" in "science fiction," it is important to point out that when most people hear the words *science fiction* usually the first thing they think about is space, rockets, and lasers. That is okay (and I like it very, very much), but science fiction is oh, so much more!

For example, really good science fiction is entertaining and at the same time intellectually stimulating. There is a long tradition of this represented by the so-called hard science fiction way of exploring the physical universe like distant planets, stars, and galaxies, but oftentimes it also explored biological and even sociological themes, even the realm of the human mind itself. One of my favorite examples of this is Isaac Asimov's *Foundation* novels. The *Foundation* novels are told from the perspective of a fictional future science, called psychohistory, invented by Dr. Hari Seldon, a mathematician and a mathematics professor. Science fiction can be inspiring to young minds, and the fictional psychohistory is no exception. Dr. Paul Krugman, a 2008 Nobelist in Economics, has credited the *Foundation* novels with stimulating his early interest in economics.[6] If we think about it, this is not surprising; economics is a highly mathematical discipline, and even if it is not psychohistory, in the words of Dr. Krugman and others, it is the closest thing we have to it.

Psychohistory was broadly based on a real scientific idea, the Boltzmann kinetic theory of gases. Boltzmann's theory essentially describes the behavior of molecules; it is able to predict the collective behavior of an ensemble of billions upon billions of molecules. However, it cannot predict the specific behavior of an individual particle. For example, using this theory, we can predict that if we open a bottle of perfume in a room, small perfume particles in gas form will diffuse through it. We perceive these particles as a smell. We can even predict how fast the smell will diffuse depending on how big the room is and on environmental conditions such as temperature, for example. However, we cannot predict which specific perfume particle will end up by the ceiling, at the floor, or up anyone's nose. In other words, we can predict the general behavior of a whole bunch of particles, but we cannot predict the behavior of an individual one.

Similarly, psychohistory was described as a highly mathematical science that could predict the collective behavior of societies (wars, revolutions, dictatorships, periods of peace and prosperity, etc.); in other words, it could predict history. However, it could not predict the actions of an individual person.

Asimov was brilliant. As my favorite Vulcan would say, "Fascinating...."
What does *Battlestar Galactica* have to do with planarians, you say? Hang on...

A great deal of hard SF deals with the concept of "life as we do not know it." This refers to life fundamentally different from our own type of life. What would be the "other" type of life most different from ours? An especially good candidate would be artificial life, like robots, for example. We have all heard about robots; they have a distinguished place in science fiction history. Robots may come in many versions, but the most extreme robots are anthropomorphic robots that are indistinguishable from real humans in terms of cognition (artificial intelligence), physical appearance, or both.

BSG excelled in this respect; the series portrayed several types of artificial life in a realistic, scientifically logical way. Overall, these characters implicitly explored the question, "What is life?" in general, but closer to home they explored the question, "What does it mean to be human?" These artificial life forms were called Cylons, and there were at least four general types in the new *BSG* universe. Two of them were humanoid types: the ones who ran the Cylon starbases and the type of Cylons who were indistinguishable from humans. The other two types were the centurions and the raiders. The centurions were loosely humanoid, "traditional" metallic robots. On the other hand, the raiders were true autonomous fighter spacecraft, where the pilots and the crafts were indeed one integrated entity.

What is the connection between *Battlestar Galactica* and planarians?

An aspect of the series that I find reminiscent of planarians is that the Cylons are, for all intent and purposes, immortal. Please remember what one of the first planarian scientists, John Graham Dalyell, said about these worms in 1814: they "…can almost be called immortal under the edge of the knife" (see chapter 6).[7]

When Cylons die in the series, their consciousness is downloaded into a brand-new body, which is stored elsewhere in a special type of vessel called the Resurrection Ship. This sounds great, but is the "resurrected" Cylon the same person? Atom by atom, we are talking about a completely different physical entity, yet they certainly feel that they are themselves, right? Before you feel smugly superior to these poor beings, guess what? Something very similar happens to you, too! Each year, about 98% of your atoms are replaced, one by one. This is part of the all-important process of self-maintenance in living organisms. This means that in a very literal sense, physically, you are not the same entity that you were, say, ten years ago, and yet, you feel that you are the same person, and of course, you are.

Or are you?

Planarians have it worse; if you cut a planarian into several pieces, each part will eventually form a complete worm; one could ask: which one of the new planarians is the original one? Remember *Planarian Man*?

I am so happy that I am not a planarian!

Also, recall what the biologist H. V. Brønsted said (we mentioned this in chapter 6): "When you have done an experiment, then go through the (scientific) literature until you have found that your experiment have [sic] already been performed by others. Only then can you be sure that you have been through the entire literature."[8]

Similarly, one of my favorite lines in the whole *BSG* series is one of its main themes: "All of this has happened before and all of it will happen again," just like Brønsted said! Or if thou art biblically inclined, "What has been is what will be, and what has been done is what will be done; there is nothing new under the sun."[9]

In science, of course, this does not exactly apply; the pace of knowledge acquisition goes faster and faster as time goes by, in part from the progress made by the invention of new technologies and ways of thinking about nature. A cursory examination of the history of science provides many examples of this. Think about relativity and quantum mechanics, two theories that account for the nature of the physical universe, from things big and very big in the former to things small and really small in the latter. The funny thing is that relativity and quantum mechanics seem to be mutually incompatible, which is rather annoying to physicists. Right there, we can see the potential for new, absolutely exciting discoveries.

What about important discoveries in chemistry and biology? Look no further than the discovery of DNA structure and the absolutely revolutionary science done in the wake of that discovery. There are many more examples, but it is time to go to the final example of a planarian and *BSG* crossover.

In 1939 R. H. Silber and V. Hamburger published a paper in the journal *Physiological Zoology*[10] about planarian regeneration.[11] In this paper they described how under certain conditions, the worms were able to regenerate multiple heads in a specific configuration called *duplicitas cruciata* (never mind what that means). Briefly, in this configuration two heads are oriented in opposite directions from each other, with two separate tails perpendicular to the heads. Figure 8.2 contains a photograph (A) and a figure (B) of a planarian in *duplicitas cruciata* form side by side with a *BSG* raider (C).

Don't tell me that the worm does not remind you of the raider!
Fasci—Er... Really cool, huh? Isn't science (and SF) wonderful?

FRINGE

Fringe is a rather popular science fiction/fantasy series that ran from 2008 to 2013. It has been compared to the *X-files* and *The Twilight Zone*. The series

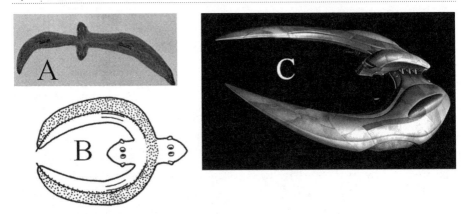

FIGURE 8.2. A, B. *Duplicitas cruciata* worm. C. Battlestar Galactica Cylon raider.
A, B from Silber RH, Hamburger V (1939). Reprinted with permission from the University of Chicago Press. C from Di Justo P, Grazier K (2010). Reprinted with Permission from John Wiley & Sons, Inc.

follows the adventures of an almost prototypical mad scientist, Dr. Walter Bishop; his equally intellectually gifted son, Peter; and an FBI agent, Olivia Dunham. This was a high-quality series that explored various truly "fringe" science topics including parallel universes, a central feature in the show's storyline.

In the episode "Momentum Deferred," aired in October 2009, Dr. Bishop tries to help Agent Dunham recover from a memory loss event by offering her a chopped planarian drink, in direct homage to McConnell's memory transfer experiments (see chapter 9). A planarian shake; why didn't I think of that?

TWILIGHT

If you have never heard about the *Twilight* books and associated movies, welcome to planet Earth! *Twilight* is one of the latest expressions of the fascination that vampires and werewolves induce in our psyche. This fascination combined with the elements of the older-than-time "boy meets girl" storyline proved to be the perfect combination to produce a media hit. To say that these books and movies are popular is a major understatement. Planarians get a mention in the movies (not in the books). I have found no information about the reason they decided to mention planarians in the story, but it is easy to speculate that they are a metaphor for immortality and physical endurance. Please recall that vampires, even before *Twilight*, are notorious for being immortal, and werewolves, although not immortal, are very hardy and especially adept at recovering from injuries at a fast rate.

THE BIG BANG THEORY

This is a hugely popular CBS comedy series centered on the lives of two physicists, Leonard and Sheldon; an astrophysicist, Raj; and Howard, an engineer. Their story is about their improbable friendship with the pretty girl next door, Penny, and other charming characters including a comic book store owner (Stuart), a microbiologist (Bernadette), and a neuroscientist (Amy Farrah Fowler, played by Mayim Bialik, who happens to actually hold a PhD in neuroscience). I must confess that I am a big fan of the series; in fact, most of my colleagues at the university are! Even though I like all the episodes, the episode that originally aired on February 21, 2013, is very close to my heart, as I received an unexpected treat: they mentioned drugged flatworms! When Amy was dealing with nicotine-addicted monkeys who were going through withdrawal in a rather aggressive way, she said "…this makes me miss my marijuana-abusing flatworms; those guys were mellow!"[12]

Now, in reality, planarians cannot exactly abuse marijuana, as they cannot smoke. Nevertheless, the effect on planarians of several examples of *cannabinoids*, the active chemical principle of marihuana plants, has been studied since the 1970s.[13,14,15]

DR. WHO

Who has not heard of *Dr. Who*? Well, I have a confession to make; even though I've heard about the series, I never watched an episode until early 2013 and now I am a *big* fan! This is the longest-running show on television. It has been around in one form or another since 1963. The story is about a humanlike alien (known only as The Doctor) belonging to a race called the Time Lords (actually, in an episode The Doctor tells someone that he is not humanlike; instead, humans look like Time Lords because they came first).

Anyway, Time Lords possess a very useful property. When mortally wounded or otherwise hurt, they can regenerate, effectively getting a new body. The catch is that they "…never know what they are going to get." In the last couple of incarnations, The Doctor has been disappointed because he is not a redhead. The Time Lords' regeneration capacities are evidently reminiscent of planarians, but there is an episode in which this is explicitly stated. In the episode "Journey's End," originally aired on July 5, 2008, the severed hand of The Doctor generates another one. Donna Noble, one of The Doctor's companions and a very colorful character by herself, is rather understandably surprised. She says: "Is that what Time Lords do? Lop a bit off, grow another one? You're like worms!"[16]

I love my science fiction.

In summary, these are just a handful of examples where planarians or themes inspired by planarians relate to popular culture. I am sure that there are many more, but at this time I want to conclude this short chapter with some planarian humor (Figure 8.3).

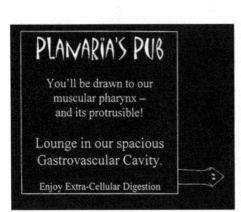

FIGURE 8.3. Planaria's Pub. Planarian birth defects.
Planaria's Pub courtesy of Dr. J. L. Wendel. Reprinted with permission from the Department of Ecology, Evolution and Organismal Biology, University of Iowa. Planarian birth defects courtesy of Alfred C. Perez. Reprinted with permission from http://judge-fred.deviantart.com.

IV

The First Brain

It is essential to understand our brains in some detail if we are to assess correctly our place in this vast and complicated universe we see all around us.

—FRANCIS CRICK

Under carefully controlled experimental circumstances, an animal will behave as it damned well pleases.

—The Harvard Law of Animal Behavior

9

THE FIRST BRAIN

EARLY, REALLY EARLY NERVOUS SYSTEMS

IN CHAPTER 3 we saw that many of the physiological mechanisms and many of the chemicals found in the nervous systems of complex multicellular organisms are also present in simpler, even unicellular organisms. We also saw that behavior, as defined by the capacity of an organism to react to its environment, does not even require a nervous system that we can recognize as such, as in the cases of plants and microbes. However, as far as we can tell, actual thinking can only appear in the context of a nervous system. Therefore, from this perspective, it seems that multicellularity is an absolute requirement to make an actual nervous system work. One of the very first questions that comes to mind is: how and when did cells learn to cooperate with each other? The answer to the "when" question seems to be "a long time ago." Bacteria are capable of intercellular communication, and bacteria and archaea are widely considered the oldest types of life on our planet (see chapter 2). As for the "how," at least in bacteria, cooperation seems to be in great part due to a mechanism called *quorum sensing* (QS).[1] For more than twenty years we have known that quorum sensing allows many species of bacteria to communicate; interestingly, this phenomenon was first discovered in luminescent bacteria that live in a symbiotic relationship with "higher" organisms like squid.[2] Quorum sensing is also involved in the development of *biofilms*, which is widely thought of as a kind of bacterial social behavior. This is not a simple biological curiosity, as QS is present in pathogenic (disease-causing) and nonpathogenic bacterial interactions with humans. It is not surprising, then, that quorum sensing is under intensive study to possibly discover and develop new antibiotic medicines, which are badly needed in light of the emergence of antibiotic resistance in multiple types of disease-causing bacteria, as well as to elucidate the myriad effects of beneficial bacteria in our bodies.[3]

As remarkable as quorum sensing is, bacteria are still unicellular. Bacteria are also prokaryotes, so we need to look elsewhere to find clues to the multicellular origins of eukaryotic organisms.

A very interesting type of eukaryotic organism, collectively called *slime molds*, has evolved a rather clever survival strategy.[4] In times of abundance, when food sources are plentiful, slime molds exist as unicellular organisms, rather similar to the amoebas that are so familiar to a student in a biology teaching laboratory. On the other hand, when food is scarce, the individual cells associate and develop into a multicellular structure called a *fruiting body* (sometimes called a slug), which has two parts, a stalk and a spore-forming portion that is used for reproduction. This relatively sophisticated type of behavior is only made possible by the highly regulated communication between their cells; they use a version of the same type of communication that exists between our cells.[5] Slime mold cells are individual, independent creatures, and yet they are capable of forming a "collective" that looks and acts as a multicellular creature. They unconsciously cooperate to reach a goal, namely, survival. Their cells take different roles, as in the phenomenon of differentiation. The cell-to-cell signaling phenomena responsible for such behavior can be seen as a forerunner to the nervous system, and it is very easy to think of scenarios that resulted in organized neuronal architecture, with actual brains as the eventual result.

Another rather interesting trait of slime molds and related organisms is that they are capable of rather impressive feats traditionally thought to be limited to "higher" animals. These include behaviors like problem-solving skills and the ability to learn.[6,7] Amazingly, they also display the ability to anticipate environmental changes based on prior experience.[8]

Still, just like bacteria, the amoebalike cells in slime molds do not possess an actual animal-like nervous system. If we think about it, from our admittedly biased perspective, the absence of a nervous system makes the behavioral repertoire of bacteria and slime molds even more astonishing. Curious creature as I am, this makes me ask myself, wouldn't it be interesting to find out whether bacterial populations have some form of self-awareness? Would it be possible at some fundamental level that slime molds, well, wonder? These are very interesting questions that science cannot answer yet.

Back to the matter at hand, phylogenetically, we begin to see the initial organization of nervous system–like structures in simple multicellular animals like sponges. Sponges are widely considered to represent the earliest example of metazoans (multicellular animals) as they have been around for close to 600 million years (Plate IV). The cells in these organisms, despite not having excitable properties as such or even true synaptic connections (see chapter 3), are nonetheless capable of cell-to-cell electrical communication via cytoplasmic bridges between cells.[9] Also, neurotransmitters like glutamate and gamma-aminobutyric acid (GABA), as well as the atypical neurotransmitter nitric oxide among others, not only exist in sponges but also, surprisingly, exhibit vertebratelike pharmacology.[10,11] Additionally, sponges react to a variety of substances that were thought to act only on more "advanced" organisms. These substances include strychnine, atropine, and the abused drug cocaine.[12] Also, there is some evidence of true nerve cells in certain sponge species. That being said, to avoid confusing the sponges' pseudo–nervous systems with "true" nervous systems, scientists sometimes talk about *neuroid systems* instead.[13]

A true nervous system seems to have appeared with the cnidarians, the type of organisms represented by hydra and various types of jellyfish (see chapter 7). The nervous system in most cnidarians consists of a distributed (oftentimes deceptively described as "diffuse," Figure 9.1) nerve net with no evident concentration of neurons centrally organized.

Regardless of the seeming simplicity of their nervous systems, many cnidarians of the jellyfish family also display one or even two nerve rings through which signals converge and which have more than a passing functional resemblance to central nervous systems. In fact, cnidarian nerve rings are widely considered to possess a sophisticated level of centralization.[14] Despite its apparent (and only apparent, as we will see) lack of centrality, the cnidarian nervous system is fully capable of cell-to-cell communication by true chemical and electrical synapses in addition to cytoplasmic bridges similar to the ones described in sponges. In hydra we also begin to see examples of directional signaling in the sense that signals travel in one direction along the neuron. In cnidarians like jellyfish we also find "pacemaker" neurons that produce the rhythms necessary for swimming, for example. It is tempting to think that since these animals do not display a "true" central nervous system, their behaviors will be rather limited. Well, yes and no; cnidarians are capable of rather complex behaviors even though learning has not been conclusively demonstrated in these organsims. For example, sea anemones and hermit crabs can exhibit a symbiotic relationship in which a sea anemone is frequently found attached to the hermit crab's shell. This is clearly beneficial for both organisms in the sense that the anemone helps deter

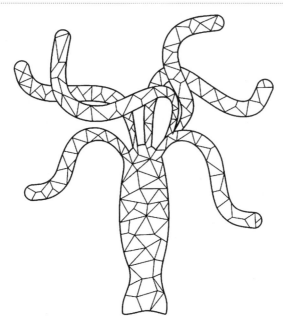

FIGURE 9.1. Representation of a hydra's diffuse nervous system. Illustration by A. G. Pagán.

predators by virtue of its toxins and, in turn, the anemone is carried around by the crab and also gets to eat from the crab's leftovers. It is unclear whether it is the crab or the sea anemone that initiates the relationship though. Nonetheless, sea anemones can react to predation and show escape behaviors like detaching themselves from the crab's shell and awkwardly swimming (more like wiggling) away. Another example of "purposeful" behavior is that jellyfish and sea anemones capture prey in very much the same way that hydra actively reaches for food with its tentacles, as Trembley observed (see chapter 7). This is certainly an indication of their perception of foodstuff in their environment.

Cnidarian nervous systems help them survive; there's no doubt about that. They also may represent an alternate type of architecture as far as central nervous systems are concerned. However, these types of organisms are not quite hunters; they capture prey that happens to cross paths with them. In other words, cnidarians do not seem to actively pursue their prey. To see nature's first hunters, we have to explore the next step in the development of nervous systems.

THE FIRST HUNTERS

All organisms must be able to capture energy from the environment so they can stay alive (see chapters 1 and 2). Evolution has come up with many strategies that allow living beings to acquire the necessary energy for survival. We are all familiar with photosynthesis and related processes, in which organisms are able to use radiant energy directly to power up their physiology. As we saw in chapter 2, we call such organisms *autotrophs*, and they are mainly represented by green plants. On the other hand, there are living beings that must consume other organisms to obtain energy; we call them *heterotrophs*. Most interestingly, there are organisms that are able to take advantage of both strategies; there's no biological law that says otherwise. For example, insectivorous plants can photosynthesize yet have developed a wide variety of strategies to capture insects. Some of the best-known types of carnivorous plants include the Venus flytrap (see chapter 3), which captures insects by trapping them between two modified leaves that rapidly close over them. Other carnivorous plants use more passive strategies. Nevertheless, again, all carnivorous plants are also able to photosynthesize, especially when cultivated with plenty of nutrients in the soil. Therefore, in contrast with purely photosynthesizing plants, in the case of carnivorous plants, the distinction between an autotroph and a heterotroph is not as clear cut.

Even more remarkably, there are several types of organisms that steal photosynthetic capabilities. This phenomenon is common enough to warrant a term of its own, *kleptoplasty*. Kleptoplasty involves the internalization of chloroplasts when a heterotroph eats a green plant; chloroplasts are the cell organelles responsible for photosynthesis (see chapter 2). In other words, kleptoplastic organisms steal chloroplasts. There are several types of life forms that use this strategy, but one of the most notable is a type of sea slug (*Elysia chlorotica*), nicknamed the "solar-powered slug" (look it up; it is really a remarkable critter).[15] Furthermore, as if this were not interesting enough, there is a class of insects that have developed the capacity of photosynthesizing without having to steal anything from plants; they do have the genes for it![16]

Let's go further; if for some mysterious reason you are not impressed by photosynthesizing marine or terrestrial invertebrates, it just so happens that it seems that a vertebrate, namely, a salamander species (*Ambystoma maculatum*), and a green algae (*Oophila amblystomatis*) live in a symbiotic relationship that allows the salamander to be an actual photosynthesizing vertebrate.[17] I would not be the least surprised if more examples of this kind of relationship are discovered in nature.

The biological world is fascinating, isn't it?

In addition to these "photosynthesizing types," in terms of relatively large animals, we are all familiar with the terms *herbivore* (plant eating), *carnivore* (meat eating), and *omnivore* ("the guys that will eat whatever they can get"). We humans are omnivores. Of these three, arguably the herbivores spend the least effort to feed, as they just have to go to the plant and eat it; after all, the plant is not going anywhere. This is not to say that plants are absolutely defenseless—far from it. As we saw in chapter 5, we know that many types of plants have evolved a series of traits to combat predation by herbivores; these strategies include structures like thorns and other physical means, as well as a wide variety of toxic or irritating chemicals, for example. Anyway, as far as omnivores are concerned, they tend to be opportunistic feeders; they usually forage for plants, but if they can catch a small animal they will gladly eat it. Additionally, some carnivores, like spiders, for example, are able to set traps for their intended prey and display quite complex behaviors for feeding purposes, but in general, they do not actively pursue their prey.

Then there are the hunters. When we think about hunters, we think about animals that actively track and seek prey using various strategies. Social hunters like lions (the only social big cats), killer whales, and wolves, among others, cooperate with each other to catch prey. Solitary hunters such as the other big cats like tigers, leopards, and so forth also make use of rather sophisticated behaviors to track, stalk, and trap prey; you see, trapping prey requires *planning*. In general, predators tend to be smarter than herbivores since they have to come up with relatively intricate behaviors to outsmart their targets. In many cases, they evidently show the capacity to plan ahead and strategize to capture prey. It is widely thought that predatory lifestyles were a result of evolutionary pressures related to the evolution of intelligence. In other words, if you want to hunt, you need to develop strategies to capture your prey, especially if your prey is faster or bigger (and sometimes both) than you or has any other kind of defensive mechanism. This is not the place to adequately explore the immense variability of predatory strategies; this would warrant a book of its own for virtually every type of such strategies, from sharp claws and brute strength to more subtle strategies like venom-delivering fangs. That said, in general terms and as far as we can tell, a well-developed nervous system is a necessary condition for the intelligent behavior that we associate with "higher" animals.

It is important to point out that a central nervous system is not strictly equivalent to a brain. True enough, in some cases, the distinction between the two is self-evident. However, sooner or later we will find a biological entity that defies a clear definition, as in the example of cnidarians. Some researchers even argue that the development of nervous system centralization is a separate process from the development of complex brains.[18] We have already seen how difficult it is to precisely define the term *brain* (see chapter 4). Because of this inherent difficulty, it is probably a good strategy to keep our options open and define a "brain" according to whatever specific criteria we are working with at any particular moment (i.e., anatomy vs. function and so on; see chapter 4). To have a rather flexible working definition of a brain is necessary but not surprising, since we have little undisputed information about the evolutionary origins of central nervous systems; thus, it is rather challenging to obtain an all-encompassing definition. On the other hand, if we limit ourselves to purely functional considerations, we can articulate a more precise characterization. From this perspective, in essence, a central nervous system, a "brain" if you will, is a well-defined anatomical region composed of neurons and allied cell types that combine and interpret multiple signals from the environment, as detected and relayed by peripheral sensory receptors. In turn, the combination of such signals usually results in responses of various degrees of complexity. These responses can be voluntary or involuntary depending on the specific nature of the processed signal, but at any rate, such responses must involve the whole body; these responses will invariably result in specific behaviors.

Some of the key anatomical elements that are implicit in this definition are the concepts of *centralization* and *cephalization*. These two terms are often used interchangeably, but they actually refer to closely related yet distinguishable properties of an animal. Centralization relates mainly to nervous system anatomical architecture. We can think of centralization as the organized collection of clusters of neurons and other cells in a direct structural association with sensory organs. This association frequently but not always defines the anterior part of the organism as a distinct structure called the head—hence the confusion between these two terms, as the actual formation of a head is known as cephalization. The cephalization process is part of a more general phenomenon in animal development, namely, the formation of the anterior-posterior axis (also known as the A-P axis). This establishment of an A-P axis essentially creates the anatomical division of an animal into well-defined "front" and "rear" parts. It is a well-established fact that the A-P axis appears very early in animal evolution and, as we have said before, any conserved trait in an organism is bound to be important for survival. Cephalization and centralization are no exceptions.

The exact point when this anatomical organization appeared for the first time in the history of earth is currently a point of much debate, as little is actually known about the evolution of central nervous systems. However, we can divide the main arguments into two main camps: a single origin versus several independent origins of centralization.[19] That said, regardless of which camp is correct (and I suspect that there is some truth in both interpretations depending on the specific point of view), the fact is that centralization happened, and without a doubt it triggered a series of events that resulted in behaviors we currently classify as intelligence, sentience, and even consciousness (human-style or otherwise).

As we have seen, in general, if you are a grazing animal, you do not have to think much to feed yourself; you would just eat any easily obtainable food. In contrast, you will need a higher degree of behavioral sophistication if you are to become a hunter. Our good friends the flatworms have been proposed as a model of what the first hunting organisms looked and behaved like. All species of planarians described so far are predators, cannibals even. Therefore, flatworms in general have been called the "first hunters,"[20] and true enough, many types of flatworms display the kinds of behaviors that merit, at least to some extent, calling them predators (see chapter 6 for a description of how planarians hunt and other relatively sophisticated behaviors these little guys display). Please keep in mind that it can be argued that a required condition for actually defining a behavior as hunting behavior is planning and therefore purposiveness. So far, we have no evidence of this type of cognitive process in planarians.[21]

That being said, as with any animals, all of the behaviors planarians are able to show are possible courtesy of their brains. The central thesis of this book is to review and expand the argument in favor of considering the flatworm brain in general and the planarian brain in particular as the first example of a brain, but not just any brain. In this sense, to be able to define what "the first brain" was is a little like defining what the first human was. It depends on how we wish to define the term *brain*, which, as we have seen, can be done in various ways.

The type of brain I am talking about is something very much like the vertebratelike brain; this type of brain possesses a series of particular features that provide a minimum of complexity at various levels. These levels include but are not limited to aspects of anatomy and cell morphology, as well as relatively complex physiological, biochemical, and pharmacological mechanisms. When you think about it, it is not that far-fetched to consider planarian brains as scaled-down versions of brains usually associated with "higher" organisms. I first saw this idea explicitly expressed in a paper published in 1985 by Dr. Harvey Sarnat and Dr. Martin Netsky, which explored the concept of the planarian brain as the "ancestor" of vertebrate brains. This paper made me think of planarians as animal models in terms of neurobiology.[22] Since then, a large body of reported research has explored the evolutionary place of platyhelminths in the grand scheme of animal phylogeny, and planarians do not seem to be in the direct lineage that resulted in us. Also, planarians are no longer considered as "basal" (see chapter 7) as they were before, a reflection of their rather complex morphology and physiology.

These facts notwithstanding, there is a series of structural and physiological features of flatworm brains that are remarkably similar to more "advanced" brains. This is part of the vibrant nature of science and undoubtedly part of its charm; as usual in these cases, prior concepts that were generally accepted are modified in light of the newly acquired information. These modified/updated concepts include aspects of the relationships between animal groups. More importantly (from the perspective of this book), in the last fifteen to twenty years the study of planarians from a neuroscience standpoint has developed into one of the most dynamic areas of research. Due to the wide variety of powerful techniques that allow us to examine the nervous system from biophysics to behavior, the humble flatworm's brain has proven an invaluable model to study many fundamental neurobiological concepts. Sometimes due to the fully warranted excitement of scientific discovery, especially in the molecular sciences, we tend to lose sight of the fact that a whole organism is made of many "molecular aspects" that act together to make things work.[23] Please note that any research on planarians does not solely pertain to invertebrate neurobiology. A significant fraction of the information that can be obtained by studying this animal model will surely find (and already has) general applications in neurobiology. For example, as we will see in the next chapter, planarians are also intensively studied from a pharmacological point of view, specifically in the field of abused drug behavior. Not surprisingly, this pharmacological research has proven to display close parallels with mammalian pharmacology, but I am getting ahead of myself. Let's explore some of the aspects that make the planarian nervous system so fascinating and relevant to modern neurobiology.

THE FIRST BRAIN

All vertebrate animals, including humans, have a nervous system that is traditionally divided into two main regions. As we saw in chapter 4 (Figure 4.1), one of these regions is named the *central nervous system* (CNS), which includes the brain and a dorsal spinal cord. *Dorsal* means "in the back" (where you would usually carry a backpack), as opposed to *ventral*, which means "in the front" (where you would normally carry a baby, for example). The second region is called the *peripheral nervous system* (PNS), which includes anything that is not the CNS (Figure 4.1). The peripheral nervous system is essentially charged with the task of relaying sensory signals to and from the CNS, as well as relaying activating signals from the CNS to effector organs (the ones that do the actual work) like muscle, endocrine glands, and so forth.

In general, the flatworms represent the earliest example of animals that exhibit definitively well-organized structural features in their nervous systems.[24] There are various schemes or *orthogonal schemes* (orthogonal refers to perpendicular lines) with various degrees of organization. With respect to flatworms, the *orthogon* organizes into a series of neural cords distinct from the actual brain, in essence, what in vertebrates would be called a PNS (Plate V). Regardless of the phylogenetic classification of the many types of flatworms in nature, the main unifying feature in all of these arrangements seems to be a concentration of neurons that defines the anterior part of the organism (as in cephalization), connected to peripheral neurons distributed throughout the body. This arrangement appears in various degrees of complexity, from the diffuse, with merely a hint of organization, to the much more clearly defined nervous system possessed by planarians, unmistakably divided into central and peripheral components. At any rate this anterior concentration of neurons is usually referred to as a "brain." Please recall from chapter 6 that even though originally all flatwormlike critters were classified as *Platyhelminthes*, more recently several types of flatwormlike animals like the acoels (see chapter 6) are not taxonomically classified as platyhelminths anymore. It is precisely these acoels and related worms that are considered truly early examples of organisms displaying a protobrain in the sense that we are talking about (Plate V).

That said, from our perspective the current scientific consensus is that among the flatworms, the planarian displays one of the first undisputed examples of a rather organized, centralized nervous system. Additionally, in evolutionary terms, planarians were widely considered one of the simplest examples of organisms that display cephalization (but see the previously mentioned acoels), a bilateral body structure, and a well-defined central nervous system. Additionally, there is evidence indicating that the organization of the planarian's central nervous system along its anterior-posterior axis displays close similarities with the developing vertebrate nervous system.[25] All of this is further evidence for the relevance of planarians to vertebrate neurobiology.

Please recall from chapter 3 that one of the characteristics of neurons is that they look very similar regardless of their organism of origin. Despite this fact, there are some subtle distinguishing differences between neuron types indeed; therefore, you would think that in terms of morphology (anatomy) and physiology (function), planarian neurons will look more like other invertebrates' neurons, right? However, this is not the case; interestingly, the planarian brain has many traits in common with the brains of more "advanced" organisms, ourselves included. For example, upon close examination, planarian neurons look more like neurons of vertebrate origin than neurons from other invertebrates (like insects, for example). One of the key anatomical features that planarian neurons share with vertebrates but not with most invertebrate neurons is the presence of *dendritic spines*, small button- or mushroomlike structures that are established players in memory and learning processes.[26] Just by looking at this fact, there seems to be something peculiar about the planarian nervous system, as all vertebrates but few invertebrates possess dendritic spines.[27] Planarians also use many of the major neurotransmitter molecules found in mammals, including humans. You may remember that neurotransmitters are chemicals that control many of the functions of our nervous systems (see chapter 3). Planarian neurons also show a fair amount of complexity in the synaptic vesicles that store these neurotransmitters and related molecules, in terms of both the variety of substances stored and their concentration within these vesicles.

The planarian brain itself is an arrowlike structure, sometimes referred to as an "inverted-U structure" (Figure 9.2), composed of three main general regions: the visual system, the lateral branches, and the spongy regions.

Briefly, under the microscope, a significant part of the planarian's brain displays a "spongy" appearance. Its central nervous system is organized into two well-defined lobes, reminiscent of the vertebrate bilobar brain. In turn, each of the planarian brain lobes branches into several lateral extensions distinct from the aforementioned spongy regions. The best-characterized planarian brain is the one from *Dugesia japonica*, which possesses nine such lateral branches. These subdivisions are strongly evocative of the organization of the twelve human cranial nerves (Figure 9.3). This does not imply in any way a direct evolutionary relationship, but it is an interesting similarity nonetheless.

Also, planarians possess a well-defined optic chiasm to which its optic nerves converge, again, in a way rather analogous to similar structures in the vertebrate brain. As an interesting sidebar, please note that in most flatworms, including planarians, their eyes are located at the dorsal surface. Not impressed? Let's try another approach to explain this. If you were a planarian, your eyes would be located more or less at the level of the nape of your neck. Yep, you would have eyes on the back of your head! To be fair, this is kind of true in many other types of animals, but it is fun to think about nonetheless. These little wormies are a true treasure trove of wonderful and curious facts, even though in all fairness, that can be said about most types of living beings on this wonderful earth of ours.

Now, back to business.

In contrast with vertebrates, which as we saw have a single dorsal spinal cord, planarians have two ventral nerve cords, which are not mere extensions of the

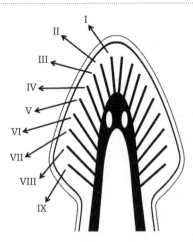

FIGURE 9.2. Representation of a typical planarian brain, indicating the nine pairs of nerve extensions. The Roman numerals are only to illustrate the planarian nerves; it is not an official designation.
Illustration by A. G. Pagán.

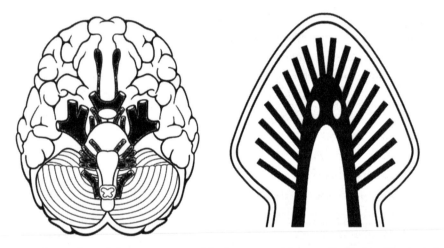

FIGURE 9.3. Comparison of the arrangement of the human cranial nerves (in black) with a typical planarian brain, indicating the nine pairs of nerve extensions.
Illustrations by A. G. Pagán.

planarian brain, but distinct structures that nonetheless connect to the actual brain—again, in close analogy with vertebrate animals. Another specific detail about these planarian nerve cords (not the lobes themselves) is that they are connected to each other by nerve fibers, giving them a ladderlike appearance.[28] This ladderlike arrangement is a highly organized example of the orthogonal structures mentioned earlier (Figure 9.4 and Plates V, VI, and VII). Figure 9.4 also shows a comparison between the central nervous systems of humans and planarians.

The First Brain | 163

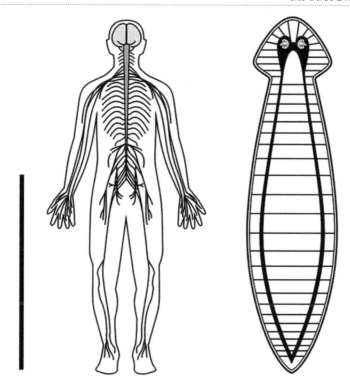

FIGURE 9.4. Comparison of the human and planarian central nervous system. The bar at left represents about 1 meter for the human drawing and 1 cm for the planarian drawing. Illustrations by A. G. Pagán.

The planarian nerve cords are not simple tubular structures; clusters of neurons (ganglia; see chapter 4) are found at regular intervals corresponding to the place where the "steps" of the ladderlike structure touch each of the two nerve cords. It is not so far-fetched to speculate about a possible role for the integration of messages traveling up or down the nerve cords.

The planarian nervous system is truly remarkable. One of the peculiar characteristics of flatworm nervous systems that does not seem to be shared by other animals is that their neurons seem to intermingle with other cell types, like muscle cells, for example; at least in some regions of their nervous systems, we can observe a combination of, say, muscle and nerve cells within the same tissue. At the present level of knowledge, the possible physiological advantage of this arrangement is unknown. Interestingly, even the planarian pharynx, the structure through which planarians eat (see chapter 6), seems to have its own distinct nerve architecture that endows the pharynx with some degree of independence when separated from the main body.

In general terms, there is quite a lot of variability in the way the nervous systems of invertebrates, particularly worms, are organized. In terms of research, this is an advantage, but at the same time, it has the potential of causing confusion. We saw this when we were talking about defining a "brain." In a similar way, different researchers come up with different ways to name the structural features they discover. Even though a thorough exploration of invertebrate nervous systems is beyond the scope of this book, there are a series of excellent review articles you may want to take a look at if interested.[29]

WHY ARE PLANARIANS AN EXCELLENT ANIMAL MODEL IN NEUROSCIENCE?

As we saw in previous chapters, planarians have proved themselves once and again as models in regeneration and developmental biology. More recently, many researchers have agreed on several aspects that make planarians an ideal animal model in neuroscience as well. There are various points that argue for considering planarians so valuable in this sense.

One of the main points is the highly probable (yet not indisputably established) single origin of a centralized nervous system before the split between invertebrates and chordates that opened the door to the evolution of vertebrates (Plate IV). Evidence for this highly apparent common origin brings about the second point, which is the structural and functional conservation of nervous system–related genes and gene products across a wide variety of species. This conservation is of course a consequence of the survival value of these gene products. Please recall that in the living world, if a particular characteristic is kept, it is bound to be important. Nature is economical; it does not keep what does not work, even if it takes millions of years to "delete" those traits not beneficial to the organism, as long as they are not deleterious on their own. Another consequence of this conservation is that in terms of pharmacology (which is essentially applied biochemistry, but do not tell anyone), we see many similarities between planarians and "higher" animals. At least to a first approximation, planarian pharmacology should be (and in fact is) closely similar to vertebrate pharmacology, for example. This aspect is further explored in chapter 10. An additional point that we have already mentioned is the close structural similarities that planarian nervous systems display with vertebrate nervous systems in structural, physiological, and developmental senses.

In addition to the multiple similarities between the planarian and vertebrate nervous systems, there is even a report of recordings of what is interpreted as a planarian electroencephalogram (EEG).[30] This is the first, but I am sure not the last, report on this type of phenomenon. Certainly more work is needed to determine whether this is a true EEG and, further, what it can tell us about the planarian brain.

From my point of view, one of the most promising areas of planarian research is the elucidation of the mystery of nervous system regeneration. This is not an abstract scientific problem; our society is plagued with neurological diseases that involve the significant loss of structure and therefore function of our nervous system in general and of our brains in particular. We do not have to look too far to see examples of this. Practically all of us have relatives with various types of these conditions, from the sadly rather common Alzheimer's and Parkinson's diseases to rarer yet devastating conditions such as Huntington's disease and amyotrophic lateral sclerosis (ALS; also known as motor neuron disease). This latter disease is particularly cruel, since in a nutshell, it progressively shuts down muscle control by the nerves while leaving the brain (and therefore the mind) relatively undamaged.[31] Then there is the problem of strokes and accidents, which are bad enough by themselves, but when they affect the brain, spinal cord, and so forth, the limited regeneration and repair capacity of our neurons makes recovery from such events rather difficult or outright impossible.

Here's where the good planarians may be able to help us. We have talked about the extraordinary capacity of planarians to rebuild the structure and organization of their brains. Even though the study of these organisms in general is fascinating enough, I believe that this is one of the characteristics of planarians that sets them apart from virtually any other animal model in neuroscience that we intend to use to learn about our own brains. Please recall that despite the relative complexity of their nervous systems, many species of planarians (not all of them though) are able to completely rebuild it, brains and all. Understandably, this is an area of research that is intensively explored by many capable and creative individuals, and with good reason. These wonderful worms are providing us with information that otherwise (at the present) we could not obtain using any other model.

Planarians have also given us information that hints at the mechanisms that essentially tell the body "the brain goes here." In 2002, a paper appeared in *Nature*[32] titled "FGFR-Related Gene *nou-darake* Restricts Brain Tissues to the Head region of Planarians," by Dr. Francesc Cebrià, a deceptively youthful scientist based at the University of Barcelona, Spain, and collaborators. FGFR stands for "fibroblast growth factor receptor," a family of proteins that in humans is associated with fundamental processes like development, embryogenesis, and cell division, among others. In terms of pathological conditions, versions of this protein are targeted for anticancer pharmacotherapies and are even related to conditions like certain types of dwarfism. The *nou-darake* gene product is most similar to the human FGFR type 1; however, *nou-darake* lacks some structural features that are present in "proper" FGRFs, hence the "FGFR-related" or "FGFR-like" ways of referring to this gene product.

In 2000, Dr. Cebrià finished his PhD with a thesis on the regeneration of muscle tissue in planarians under the supervision of Dr. Rafael Romero, of the University of Barcelona, Spain (the planarian research "hotspot" I mentioned before). His interest in the planarian nervous system took him to Japan to obtain postdoctoral training under the supervision of Dr. Kiyokazu Agata, an established researcher at Kyoto University that, according to Dr. Alejandro Sánchez Alvarado (whom we met in chapter 6), is "...the main driving force on planarian neurobiology." Dr. Cebrià evidently concurs, as he describes Dr. Agata as "one of the most enthusiastic and engaging scientists that I have met in my career." Dr. Agata's team is also very much active in the molecular characterization of neurotransmitter systems in planarians. In fact, a great deal of what we know about the planarian brain has come from Dr. Agata's research group (For a particularly aesthetic example of the planarian *Dugesia japonica*'s CNS structure from Dr. Agata's research group, see Plate VIII.)

Dr. Cebrià's original plan was to characterize at the molecular level the planarian central nervous system. Not surprisingly, one of the main factors that made planarians attractive for him was their regeneration capacities. He found regeneration "*...an amazing and striking biological process in itself...*," but further, the bare fact that something like a brain could be regenerated grabbed his imagination, and rightfully so. His original idea was to investigate the differences between the phenomena of embryogenesis and regeneration. Alas, this was not to be, due to the relative intractable nature of planarian embryogenesis. However, at the time, Dr. Agata's group was interested in characterizing head-specific genes in planarians, which presumably were directly related to their brains. They were astonished to find out the high degree of gene similarities between planarian and vertebrate brains; moreover, they also described the quite unexpected molecular complexity of the planarian nervous system.

Eventually, in addition to the *nou-darake* research, Dr. Cebrià was able to propose a logical model for planarian brain regeneration, a model that is still refined and is the starting point for practically every subsequent work in the area. He also spent several years in the laboratory of Dr. Phillip Newmark (whom we also met in chapter 6) further working on planarian CNS regeneration. This was yet another scientifically productive time of which Dr. Cebrià speaks fondly. Let's go back to the *nou-darake* story.

Nou-darake (ndk) means "brains everywhere" in Japanese. This is a "reverse hint" on what the product of this gene does. Experiments show that when this gene is inhibited through various methods, brain tissue forms throughout the body of the planarian upon amputation of the head. Now, under normal circumstances the ndk gene product is expressed in the head of the planarian and oddly enough in the pharynx. It is only when the gene is *blocked* that we will see the "brains everywhere" phenotype. The current interpretation of the mechanism of the ndk gene is that it binds and traps growth factors that induce the differentiation of neoblasts (see chapter 7) into brain tissue. These factors are produced in response to head amputation. Thus, when ndk's activity is reduced, these factors are able to diffuse to the rest of the body of the worm and trigger brain tissue formation. The first obvious question is, since ndk is expressed in the head tissue, why does it not seem to inhibit brain formation in the head? Good question. Even though the matter is not settled, the current interpretation has two parts. The first part is that these factors are more highly concentrated in the head region; after all, it is where they are produced. The second part is that the FGFRs display higher affinity for the growth factors than the ndk product does. The possible net effect is that when ndk is not blocked, the growth factors are able to bind their intended FGFR targets better than the ndk product and trigger the formation of brain tissue; any stray molecules of growth factor that escape to the rest of the body are then trapped by ndk. On the other hand, when the ndk is not blocked, these "strays" are able to bind to the neoblasts and trigger brain formation. Isn't this an incredibly interesting mechanism? As we said in chapter 5, *everything starts with binding*!

Another interesting detail about the phenomenon of planarian brain regeneration is the general manner through which the whole brain regenerates. The simplest assumption would be that upon decapitation, the new brain somehow "sprouts" from the nerve cords. However, based on the most recent studies, when a planarian is decapitated the new brain does not seem to originate from the nerve cords. Rather, brain tissue seems to originate from cells separate from the two nerve cord stumps. Along the way, the nerve cord–brain connections are re-established. This is certainly a hot area of research; in fact, these are certainly exciting times in regeneration research, with the promise of many more developments that can lead to treatments that were only dreamt of just a little while ago.

Another important question in developmental biology can be essentially stated as alternatively "Which way is up? or "Which way to the front?" Planarians are also one of the main characters in the elucidation of the genes and mechanisms responsible for this phenomenon.[33]

We have barely scratched the surface of the processes of brain regeneration and body patterning in planarians; even though we have made a lot of progress, there is still a lot of work to do for sure.

The flatworm nervous system has fascinated generations of researchers, and rightfully so. Prior to the modern explosion of available molecular techniques, the emphasis was essentially on morphology and physiology, with many talented researchers attacking the problem through multiple approaches. Later on, this information was complemented by genetics and molecular/cell biology, as well as by taxonomy and phylogenetics. All of these approaches are still ongoing. Should you want to know more details, there are plenty of resources available to explore this interesting topic in more depth.[34]

One of the most intriguing aspects of the general architecture of animal nervous systems is the relative arrangement of the nerve cord(s) and the digestive system. In vertebrates, including us, the nerve cord runs dorsally, whereas the digestive organs are oriented ventrally. In contrast, invertebrates tend to display the inverse arrangement. It is not clear what the relative advantages and disadvantages of one arrangement versus the other are. Then there is the case of octopi, squids, and related animals where their brain is wrapped around parts of their digestive system, but this is a different, albeit fascinating story in itself, which is explored in some more detail elsewhere.[35,36]

In this sense, with respect to the vertebrates and "typical" invertebrates, there are several variations of a theory that in broad terms states that the vertebrate nervous system arrangement evolved from the invertebrate version by developmentally turning "upside down." This is not as bizarre as it sounds. This seems to be a logical consequence of comparative anatomy and physiology, and this notion has been brought to light again in terms of molecular biology.[37,38]

As interesting as these aspects of nervous systems are, arguably the main "objective" of the nervous system in terms of function is the generation of behavior all the way to what we know today as psychology. Planarians are also involved in this chapter of the history of science.

VERY BRIEF COMMENTS ON PROTOPSYCHOLOGY

In most animals, the brain is largely responsible for the generation of behavior based on its interpretation of the signals it receives from the environment. In other words, in animals, behavior is a rather important consequence that results from the multiple lines of communication between all the cells in an organism, usually as a reaction to external stimuli. These communications are arranged into organized, integrated signals that are then integrated by the nervous system. In turn, the nervous system organizes and sends the appropriate responses and their reactions to environmental stimuli. Please remember that each and every living entity on our planet, bar none, displays some kind of behavior. In turn, such behaviors may or may not be evolutionarily conserved depending on their ability to increase the chances of survival and reproduction (see note 20).

At their most basic level, after everything is said and done, as far as animals are concerned, ethology (animal behavior) and psychology (which is widely considered the "ultimate" manifestation of ethology, at least from our human point of view) are essentially two aspects of applied neuroscience (see chapter 3). This includes behaviors related to learning and memory, two rather useful traits for survival in nature. At this point, you may not be surprised to learn that planarians are also central characters in the modern pursuit of the rudiments of learning and memory. In fact, for a while at least, there was a lot of enthusiasm for a field called "protopsychology," which was used in the context of planarian learning and memory. To give you an idea of how scientifically "hot" (at least for a short time) this field was, it even made the cover of the February 1963 issue of *Scientific American* and an article in the same issue by Dr. Jay B. Best, one of the pioneers in the study of behavior using our good friends the planarians (Plate IX).[39] If you are a science enthusiast, you surely know that *Scientific American* was for a long time *the* scientific magazine to read if you wanted to be properly informed in matters of science and technology. With the emergence of many other science publications with various levels of seriousness, the influence of *Scientific American* was somewhat diluted, and it had to adapt to the level of a more general magazine while retaining most of its status in the field of general science reporting.

The reason planarians began to be the favored behavioral model is that despite their relative simplicity compared to other organisms, they display a rather surprisingly sophisticated set of behaviors, including learning, memory, and other associated phenomena. Recently, they have also been found to display specific, reproducible behaviors related to the exposure of some of the very drugs that are abused by humans. We'll see some of these behaviors in the next chapter, but before that let's explore some of the basic "protopsychological" behaviors of planarians.

To the best of my knowledge, it seems that the first report on a type of planarian behavior that can be associated with learning was published in 1908,[40] which was based on a pretty simple planarian behavior. Under normal conditions, the worms glide at the bottom of the container they are in. If they perceive any kind of disturbance in the water, they momentarily stop. This paper described that planarians of the species *Dugesia gonocephala* (which is really *Dugesia japonica*)[41] and *Dugesia maculata* (which is also known as *Dugesia tigrina*) would briefly stop gliding if the container was rotated ever so slightly. Remember that planarians respond negatively to tactile stimuli (see chapter 6). Water perturbation is such a stimulus. When they notice the disturbance their reaction is to stop and "see what happens" (of course, I am personifying this behavior for the sake of clarity).

However, if the investigator kept rotating the container constantly at roughly one-second intervals, pretty soon the worms got accustomed to the water disturbance and therefore they stopped reacting to it and did not stop; they kept gliding. In essence, they paid no attention to the unexpected water movement because it was not unexpected anymore; they remembered, albeit for a short time as we'll see, that nothing dangerous happened. However, if they were left alone for only one minute, they "forgot" that the water movement did not indicate anything dangerous and recovered their normal behavior. In other words, they reacted to the stimulus again and stopped gliding when they sensed it. In strict terms, this is not learning; rather, this is called *habituation*. You are capable of that too. Let's suppose that you hear a loud noise. The first time you hear it, you may be startled, but if the noise keeps repeating, you will get used to it in time. If you are a father, I bet that you are pretty familiar with this phenomenon:

…dad, dad, dad, Dad, Dad, Dad, DAD, DAD, DAD, DAAAAAAAAAD!…
Did you hear something?

Anyway, planarians also showed habituation to other kinds of stimuli, like the light-avoidance behavior that we described in chapter 6, among other types of stimuli. Eventually, the first data demonstrating planarian learning was published in 1920 by Paul van Oye, a zoologist and physician from Belgium. To demonstrate planarian learning, among other things, he trained planarians to crawl down a wire to get food. (Food is always a great incentive for any organism to do anything. Don't believe me? Have you ever seen students eating?) Anyway, van Oye's paper was titled "About Memory in Flatworms and Other Biological Observations in These Animals."[42] This report was not generally known; in all probability it was overlooked because it was published in a relatively obscure Dutch journal. It was brought to the attention of the general community of planarian researchers by Dr. Roman Kenk (whom we met in chapter 6) in the 1960s. The van Oye maze was rapidly established as a useful method in these type of studies.[43]

In the late 1930s, other studies that explored flatworm learning were published, but the lack of adequate controls was a pervasive problem that limited the interpretation of such experiments. It was not until about 1955 when the question of planarian learning was examined again.

Let me say again that at a fundamental level, psychology is essentially behavioral neuroscience. As such, it stands to reason that the origin of many of the properties of nervous systems that give rise to behaviors that psychology can explore are found in simpler organisms. Some of these organisms include planarians. After these early twentieth century reports on planarian behavior that we just saw, in the 1950s several scientists "rekindled" the use of planarians, hoping to gain insight into the mechanisms of learning and memory. This episode of the history of science is somewhat complex, interesting yet controversial for sure, and misunderstood by most.

By all accounts, including his own, James V. McConnell was a very happy guy and therefore a very happy scientist. Time periods notwithstanding, I wish I would have had the opportunity to meet him. I suspect that at the very least, we would have had some interesting and amusing conversations. Ironically, his light-hearted nature most likely influenced the perception that other scientists had of him, and by extension, of his research. This is simply human nature. For good or bad, the way someone is perceived by others, especially peers and colleagues, influences how the same person is perceived professionally. There is no doubt that this was a factor in how McConnell was seen by the general scientific community.

McConnell joined the University of Michigan in 1956, right after obtaining his PhD in experimental psychology at the University of Texas. He was as driven and ambitious as they come and he rapidly built a reputation as an excellent scientist, academic, and educator. It is important to say right away that McConnell was no "fringe" scientist in any sense of the term. He authored many sound scientific papers and a very well-received textbook on introductory psychology. Sadly, the impression that many people in the scientific world have of him is rather different.

He started working on planarians with a classmate, Richard Thompson. At the time, both Thompson and McConnell were PhD students under the guidance of Dr. M. E. Bitterman, who originally suggested to them the use of planarians as an experimental subject.[44] Ironically, the first "hostile" critic of memory research using planarians was Bitterman himself; he considered Thompson and McConnell's experimental designs "inadequate" and their methods "subjective." Nonetheless, both Thompson and McConnell went on to distinguished careers. Like many friendships, though, these two young scientists differed in their personalities. Thompson was by far more low-key than McConnell, who was, in turn, more...flamboyant.

Prior to his career as an experimental psychologist, McConnell was at various times a radio announcer, disc jockey, and script writer. He was also a science fiction writer, and a good one too, as his work was published in specialized magazines. In fact, he used his artistic abilities to write a science fiction story titled "Where Is Yesterday?" that in a not very subtle way mocked his PhD advisor Galileo-style (the antagonist in the short story was called Sauerman, clearly a wordplay on Bitterman) and included the story as an introduction to a chapter in his psychology textbook, *Understanding Human Behavior*.[45]

There are many aspects that influenced the already complicated relationship of McConnell and science. He was essentially an innovator. While there is nothing wrong with that in itself, it is evident that in many instances the enthusiasm he showed for his ideas caused him to be perceived as someone who "wants to run before learning to walk," and in fact he tended to do that in the sense that he was described as someone who "...raced from one exciting phenomenon to the next without comprehensive experimental analyses or adequate controls."[46] That is bound to rub many a scientist the wrong way. We will return to McConnell's "scientific sins" soon, but in the meantime let's talk for a while about the things he got right.

The paper that started it all was authored by Thompson and McConnell (in that order); it was titled "Classical Conditioning in the Planarian, *Dugesia dorotocephala*" and published in 1955 in the *Journal of Comparative and Physiological Psychology*.[47] Briefly, the paper demonstrated that planarians were able to relate the perception of a certain stimulus (a flash of light) to another stimulus (a mild electric shock). In this design, the flash of light is termed a *conditioned stimulus*, because by itself (in theory) it should not induce any apparent behavior. On the other hand, the electric shock is termed an *unconditioned stimulus*, because an electric shock in itself will induce a behavioral response from the planarian (and anything else for that matter). The trick was to combine both the conditioned and unconditioned stimuli in such a way that planarians would relate the exposure of the former with the latter and react to it. They did this by flashing the light and following it up with an electric shock. In essence, planarians were trained to expect a shock right after "seeing the light." It did work, but planarians are not master learners; it took on average about 150 training sessions for a typical worm to learn this (you or I would learn this in way fewer trials). They did learn nonetheless, and that was the main point. That said, from this very first paper, the controversy started. This was the very same paper that was criticized by Dr. Bitterman.

Despite this, it seemed very clear that planarians were indeed capable of learning. The really interesting part was a series of experiments where they cut the planarians in half and then allowed each part to regenerate their missing sections. After that they tested again using the same paradigm (flash of light/electric shock). As expected, the anterior portion containing the head that had regenerated a new back portion showed evidence that they actually remembered, as they needed fewer trials to be trained to respond to the light. What was very much unexpected was that the tail portions that had regenerated a brain "remembered" as well as the original heads! In fact, other more refined studies seemed to indicate that the newly regenerated brains remembered *better* than the original brains. McConnell speculated that this was so because the regenerating brains were shielded somehow from environmental interaction—clearer heads if you will. Further down the line, other researchers demonstrated that in similar experiments if the regenerating tail sections were exposed to a solution of ribonuclease, an enzyme that degrades ribonucleic acid (RNA), these tail sections did not remember their training. This suggested a role of RNA in memory formation and retrieval in the worms.[48]

The planarian learning phenomenon seemed to be real, and even though many researchers were not able to replicate his results, many others did, sometimes under a variety of experimental conditions. That in itself was a good and a bad thing, because it tended to frustrate a good number of people. Factors that were mentioned that seemed important in these experiments were whether the worms were facing the anode or the cathode at the time of receiving the electric current, the phase of the moon (Yes, this was suggested!), the altitude above sea level, the electrical frequency of the shocks, and several others. Still, none of the arguments against McConnell's research were as devastating as the publication of a paper by two distinguished scientists, Michael Bennet, a biochemist, and Melvin Calvin, a Nobel laureate in chemistry for his work in photosynthesis. Helped by several of McConnell's former students, they tried to replicate the planarian conditioning experiments, yet they reported that no effects consistent with conditioning were observed. However, it is important to point out that Bennet and Calvin published their negative results in a serial publication called the *Neurosciences Research Program Bulletin*.[49] This was essentially a "club" publication (most likely not peer-reviewed) of the Neuroscience Research Program Organization. This was a rather elite albeit informal association no one could join unless invited, and only members seemed to be allowed to publish in their journal. I am not even sure if this association exists anymore. Not surprisingly, McConnell disagreed with Bennett and Calvin's interpretation and data analysis and wrote a rebuttal of their arguments. Alas, he was not allowed to publish his rebuttal in the same journal. He had to express his disagreement elsewhere.[50]

Then there is the matter of the phenomenon of the chemical transfer of memory. In a series of experiments that took advantage of the cannibalistic tendencies of planarians, McConnell and others claimed that if trained planarians were fed to hungry, untrained worms, the untrained worms seemed to acquire the trained worms' memory.[51] As expected, this report generated (or should I say regenerated?) more controversy.

One of the factors that most likely influenced the perception that the general scientific community had about McConnell is that he actually funded his own journal, the *Worm Runner's Digest*. This publication published a mixture of serious science with satire and humor. Understandably, this tended to confuse people, and in a couple of instances irate scientists wrote McConnell to complain that they wasted time reading an article they thought was legit, not realizing until the end that they were reading a humorous article. As expected, this kind of thing did not help the case for McConnell's scientific credibility.

That said, the undeniable truth is that planarians display something very much like learning, and that the tail seems to remember such learning. That in itself is remarkable, and yet further research in this area was essentially put on hold for a scientific generation or so due to how McConnell was seen by many of his peers. Clearly much more needs to be done in this area, and there are laboratories working on these topics at the time of this writing. There is a very exciting recent development in this matter. One of the main sources of inconsistency in behavioral experiments is the human factor. This means that in most cases, there can be an unconscious bias from the experimenter's side. This was a *big* factor in the McConnell controversy. Well, computers have come to our rescue. Experiments published in a 2013 paper[52] from the laboratory of Dr. Mike Levin at Tufts University demonstrate that indeed, planarians can learn and the tail remembers! They used an automated system[53] that eliminates personal bias through a training paradigm slightly different from previous approaches. The results of this paper are a modern partial vindication of McConnell. These are truly exciting times in planarian neuroscience. I have a feeling that this is just the beginning; it seems that McConnell was indeed on the right track as far as planarian memory is concerned. However, the matter of the chemical transfer of memory is still far from being properly explored.

The story of McConnell and planarians is much richer than I am able to describe here. This is a rather famous (or infamous, depending on the point of view) episode in the history of the neurosciences. Several accessible accounts of this story are found in the general literature and could easily make the subject of a separate book.[54,55]

Finally, let's explore how planarians are helping the field of pharmacology.

Revolutionary new drugs...
—EDWARD O. WILSON[1]

Nature is the best chemist...
—ONÉ R. PAGÁN[2]

10

FROM CORALS AND PLANTS TO PLANARIANS AND RATS

PLANARIANS IN PHARMACOLOGY: NICOTINE AND COCAINE

I REALLY HOPE that through this book I have been able to convince you that doing research with planarians not only is interesting but also has many distinct practical advantages. In addition to the unique characteristics that make planarians scientifically intriguing, working with them is technically easy and relative inexpensive; there is no need for complex incubators, big tanks, or any kind of specialized equipment to keep them in. A big plus is that they do not bite and cannot harm us in any way! Also, these organisms provide an excellent alternative when experiments using vertebrate animals are ethically, experimentally, or financially impractical.

An additional advantage of the planarian model is that they are being intensively studied from the perspective of molecular biology. As we saw in chapter 7, a planarian genome project for one planarian species (*Schmidtea mediterranea*) is pretty much completed (as far as these things can go), with two additional flatworm species including *Dugesia japonica*, first described by Dr. M. Kawakatsu and Dr. K. Ichikawa (we met Dr. Kawakatsu in chapter 6), and the marine flatworm *Macrostomun lignano* closing in really fast. This wealth of information derived from various sources, from the taxonomical, physiological, behavioral, and molecular fronts, makes the case for flatworms in general and planarians in particular as rather unique animal models. So, to summarize, these wormies:

- Possess a nervous system sharing many similarities with our own nervous system, including parallels in anatomy and cell biology, biochemistry, physiology, and pharmacology, among others
- Display a rather varied repertoire of behavioral responses, including responses to drugs abused by humans
- Can learn and remember
- Are being molecularly characterized (biochemistry and genetics)
- Are easy and practical to handle and work with
- Are rather proficient at complete regeneration of all body parts, including their nervous systems and brains

When taking these and other factors into consideration, it is not surprising that planarians are true rising stars in pharmacology research. This is especially evident in neuropharmacology and behavioral pharmacology; planarians exhibit a variety of responses and relatively complex behaviors when exposed to substances with psychoactive effects in "higher" animals like vertebrates. Furthermore, these worms display various types of curious behavioral responses when exposed to drugs typically abused by people; in these cases, such planarian behaviors are reminiscent of behaviors displayed by humans upon drug intake and abuse and might lead to a more complete understanding of these intriguing behaviors and offer insights into possible treatments to alleviate them.

The first report on abused drug behavioral effects in planarians explored the effect of morphine on dependence and tolerance.[3] It was published in 1967 by Dr. Herbert L. Needleman, a pediatrician/psychiatrist with a rather interesting story besides his one planarian paper. Quite a character![4]

Several types of abused drugs have been studied using the planarian model. Two of these drugs are the very well-known compounds cocaine and nicotine. For this reason we will use these two drug examples to illustrate planarian pharmacology, keeping in mind that other abused drugs like opiates and amphetamines have also been studied in the planarian model. We are all familiar with the effects of abuse of these drugs on our society, so we will not elaborate on those here. Rather, we'll explore the consequences of the exposure of planarians to cocaine and nicotine since the planarian model has the potential of shedding light on the effects of such drugs in humans. It is increasingly evident that the study of such drugs in planarians has the potential of offering insights that may be applicable, at least in principle, to human biology.

For example, if you expose planarians to nicotine, depending on the specific conditions, you will observe a variety of concentration-dependent responses. The first apparent effect is a decrease in motility (which may remind you of a human relaxing when smoking). This "relaxing" effect of nicotine on planarians has been known for some time and was taken advantage of when researchers needed to slow down the worms. In essence, researchers used nicotine to narcotize them.[5,6] More recently, planarians have been found to display seizurelike behaviors when exposed to high nicotine concentrations[7] (nicotine and, in fact, any and all drugs can be toxic to humans, too).

The expression of these nicotine-induced behaviors is being explored more systematically for research purposes, with the main objective of screening for substances capable of preventing or alleviating the effects of nicotine in biological systems—ostensibly down the line, in humans. When the drug is given to the worms in a relatively high amount over a rather short time, this mimics the situation of acute drug exposure or even drug toxicity. On the other hand, when the worms are exposed to the drugs in relatively lower amounts over a longer time, this mimics a situation of chronic exposure. In this case, when drug intake is stopped, an altogether different type of behavior may appear, namely, *withdrawal-like behaviors*. These are very well known and characterized in people.

Planarians have proven surprisingly useful to study withdrawal-like behaviors. This was actually quite unexpected. It was generally assumed that these withdrawal-like responses required the more sophisticated vertebratelike neuronal architecture. Remarkably, if you give nicotine to planarians for a certain period of time and then take the nicotine away, they behave as if they were addicted to nicotine, as inferred by their withdrawal-like behaviors! That begs the question: how in this good earth can you tell whether a planarian is addicted to nicotine? Well, to start with, they squirm and twitch. In humans such behaviors would be interpreted as "nervousness."[8] Other stereotypical behaviors observed in planarians are shown in Figure 10.1. These behaviors, while admittedly arbitrary, are quite reproducible and are an integral part of a paradigm that uses planarians as an invertebrate model of withdrawal responses.

This withdrawal behavior model was originally proposed and described in planarians exposed to cocaine, and it has proven useful to test a wide variety of psychoactive substances.[9] These types of behaviors also occur when planarians are exposed to other addictive substances like opiates and amphetamines, for example. Evidently, these interesting drug-induced behaviors of planarians make them intriguing and useful models of drug abuse. In a practical sense, the beauty of this is that if we can characterize and reproduce withdrawal-like behaviors in a consistent manner, we can use that as a model to screen for potential substances

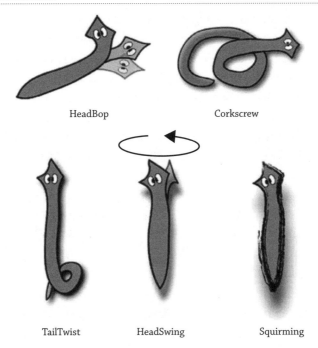

FIGURE 10.1. Some planarian withdrawal-like behaviors.
From Raffa RB, Desai P (2005) Description and quantification of cocaine withdrawal signs in planaria. Brain Res. 1032(1–2):200–2. Copyright © Elsevier. Used with permission.

that may counteract such drug effects in the worms. The next thing that happens after drug candidates are identified based on these studies is to test them using similar paradigms in "higher" organisms. We will see an example of this soon.

The earliest reference I have been able to find on the application of cocaine in research using planaria was published in 1891 in the journal *The American Naturalist*, where they described the use of cocaine to immobilize the worms for microscopical studies, just as nicotine was previously used for.[10] The next report appeared in 1907, where cocaine was reported to be used to narcotize a new planarian species discovered in Hawaii.[11] In these two works, even though cocaine *was used* in planarians, they did not *study* cocaine. The drug was merely used as an anesthetic agent.

THE BEGINNINGS OF SYSTEMATIC PLANARIAN PHARMACOLOGY RESEARCH

Apart from the use of cocaine as an anesthetic agent to slow down planarians, as far as I am aware, no other cocaine/planaria research reports were published until two papers appeared in 1996–1997 authored by a group of physicians who specialized in neurology that hypothesized that planarians would make useful models in neurobiology. This group was led by Dr. Guido Palladini of the University of Rome, Italy. Of the two cocaine/planarian papers, one had a pharmacology/behavioral emphasis while the other had an anatomical/morphological point of view.[12,13] In these two papers, they described the relationship between the planarian dopaminergic system and cocaine and its parallels to vertebrate systems using behavioral experiments. They also described an apparent neurotoxic effect of cocaine in terms of changes in neuronal morphology, as explored by histology and electron microscopy. It should be no surprise that, as usual, there is a human side to the story.

Dr. Guido Palladini and Dr. Antonio Carolei pioneered the modern era of the systematic use of planarians in pharmacology, as explicitly proposed in the early 1970s. The paper that started it all was titled "Proposal of a New Model with Dopaminergic-Cholinergic Interactions for Neuropharmacological Investigations" (1975).[14]

As Dr. Carolei told me in an email communication, he and Dr. Palladini were having a scientific conversation that somehow turned into a philosophical discussion on immortality (in my imaginative mind, I'd like to think that this conversation happened while imbibing fine Italian wine). At one point Dr. Palladini remarked that of all known animals, planarians seemed to be one of a very select group of animals that had the potential of becoming immortal by virtue of their famous regeneration capabilities. According to Dr. Carolei, it was in the course of that very conversation that his interest for planarians began. His main motivation to consider planarians as a neurological animal model was that they showed a unique potential to be developed as an important model for neurological conditions, specifically Parkinson's disease, because it was already known that planarians used the neurotransmitters dopamine and acetylcholine. Additionally, the structural features in the planarian nervous system that reminded them of the organization of vertebrate nervous systems made the worms attractive to them.

Drs. Palladini and Carolei, together with Drs. Giorgio Venturini, Vito Margotta, and Francesca Buttarelli, among other colleagues, began to explore planarians as animal models for neurological conditions and published quite a few papers on the topic; many of their papers also dealt with pharmacological aspects. Again, not surprisingly, their main reason for using these flatworms was to take advantage of their simpler yet vertebrate-relevant nervous system; for this purpose they chose a readily available species of planarian, *Dugesia gonocephala* (which is the same as *D. japonica*),[15] a species similar to *D. tigrina* and *D. dorotocephala*.

Overall, several members of Dr. Palladini's original group or their academic offspring published close to forty papers or book chapters on planarian neurobiology and pharmacology between 1975 and 2008, including investigations on their fundamental biology in terms of the neurotransmitters acetylcholine, dopamine, neuropeptides, and nitric oxide, among others. Also, they investigated the effects of drugs like cocaine, cannabinoids, and opioids, among other compounds, using the planarian model. Their insights on the applicability of planarians as surrogates for more complex nervous systems, specifically in terms of behavioral pharmacology, are summarized in a recent review.[16] In this review, they explicitly stated the close parallels between the pharmacology of planarian and vertebrates, specifically mammals.

THE TEMPLE UNIVERSITY TEAM

In 2001, a paper describing behaviors reminiscent of "withdrawal symptoms" upon exposure to cocaine was published by a group at Temple University in Philadelphia, Pennsylvania, led by Dr. Robert Raffa. The paper was titled "Cocaine Withdrawal in Planaria."[17] The publication of his paper was likely a key factor in the renewed interest in planarians as a pharmacological model. Dr. Raffa began to consider planarians as a pharmacology research subject at the suggestion of Dr. Timothy Shickley, a PhD student in his laboratory. However, Dr. Raffa had a previous connection with planarians; even though he was originally trained as a chemical engineer, he also earned an undergraduate degree in physiological psychology. As part of his psychology studies, he chose to write a paper on the McConnell experiments and controversy that we saw in chapter 9.

In 2002, Dr. Raffa joined forces with an up-and-coming young pharmacologist, Dr. Scott Rawls. They, alongside their students, enthusiastically restarted the planarian experimental pharmacology field. From then on, several other research groups with an explicit interest in planarian pharmacology began to work in the field, including other laboratories in the United States, as well as research groups in Japan, the United Kingdom, Italy, and Spain and my own research group, among others. In particular, it is important to highlight the work of Dr. Kiyokazu Agata of Kyoto University and his collaborators. We mentioned the seminal contribution that Dr. Agata's team had to the study of the planarian brain in chapter 9. His laboratory is also very much active in the molecular characterization of neurotransmitter systems in planarians, as well as in planarians in toxicology. Please remember that toxicology is pharmacology's sister science, for which planarians serve as useful animal subjects, with all the advantages described previously and more. Overall, the field of planarian pharmacology and toxicology has steadily grown over time, as illustrated in Figure 10.2) and it shows no sign of slowing down.

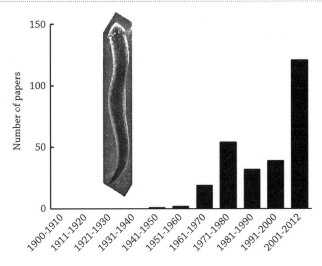

FIGURE 10.2. Planarian papers on pharmacology or toxicology. The graph was generated through the PubMed website (http://www.ncbi.nlm.nih.gov/pubmed/) using the following search: (planaria or planarian or planarians) and (pharmacology or pharmacological or toxicology or toxicological).
The picture of the planarian was taken by the late Dr. Dick Woodruff, Emeritus Professor of Biology, West Chester University.

Now I would like to tell you the story of how I found my way to planarian research.

THE JOY OF DISCOVERY

As I said in the introduction, I never imagined that I would be working with flatworms. Until a few years ago I actually knew very little about them besides the basics: they were flat, they were worms, and if you cut them up, they would regrow whatever they lost.

Flash forward to graduate school.

In 2001, I began working on my PhD degree under the supervision of Professor George Paul Hess, of the Department of Molecular Biology and Genetics at Cornell University.[18] My project involved the biochemical pharmacology of certain proteins within nervous systems that are the main targets of abused drugs like cocaine. I was in a unique situation. George's group is mainly about biophysical chemistry, with the "bio" part just added (I suspect) for its "sex appeal" for grant application purposes. I was primarily a biologist, while virtually everyone else in the laboratory was a chemist or a chemical engineer. George himself (at

Cornell it is traditional to call professors by their first names)[19] is a highly mathematical biochemist/physical chemist. Therefore, I liked thinking of myself as the "resident biologist" of the Hess Lab. The specific topic of my dissertation was the biochemical pharmacology of the dopamine transporter,[20] a neuronal protein that is in charge of the termination of dopamine-controlled neurotransmission. This protein is the pharmacological target that is related to cocaine's behavioral effects in humans.[21]

Between 2002 and 2004, as my dissertation was beginning to take its final form, a paper in a reference search caught my eye. It was Dr. Raffa's 2001 paper, the one that we saw previously on planarian withdrawal behaviors. I printed it and immediately went to see George, excitedly telling him that we should try planarians in the lab. George was standing; at 6 foot 4 inches, he easily towered over me (and no, I will not tell you by how much). He then said something like, "Well, when you have your own laboratory, you can play with these worms." And the rest is history; that's exactly what I did! Incidentally, I think the last time we talked, I reminded him of this conversation (yes, I am an evil, evil man). And you already know what I have been doing professionally since then. In an unexpected turn, I ended up as a faculty member at a university near Philadelphia, Pennsylvania, and as the saying goes, one thing led to another and now I collaborate with Dr. Raffa, whom I now call Bób (yes, with an accent over the "ó"; long story). We recently published the first collaborative paper of the Pagán, Raffa, and Rawls laboratories.[22]

This section is a little different from the previous ones. Here I narrate the story of some of the research that I have experienced firsthand, leading up to my research using planarians. This started with my master's and PhD research, and later as the head of my own laboratory. Along the way, I have had the pleasure and privilege of establishing friendships and research collaborations with some of the scientists who are leading the modern use of planarians as an animal model, especially in pharmacological research. Two of these researchers are Drs. Raffa and Rawls, as I said earlier.

To be completely honest, I was actually very reluctant to write about my own role in the planarian pharmacology field. I did not want to sound self-promoting or anything like that. Bób Raffa convincingly argued that this was not the case, as the bare facts were that I loved science in general and pharmacology and neurobiology in particular, that my story undoubtedly illustrated the "unexpected connections" aspect of scientific research, and that I did the work after all. I think he was right. Another reason I am doing this is that I like talking about my research with anybody within earshot. As Carl Sagan once said about science, "…when you are in love, you want to tell the world."

So, here it goes; this is how I got involved in planarian research.

There is nothing quite like the scientific process. There is something remarkable about examining the multiple connections that ultimately result in a discovery. In many cases, scientists have no idea where their observations will lead them until they are very close to the actual implications of such findings. Then, of course, they invariably feel that the answer was in front of them all along, and inevitably many more questions come up after the initial fact. My case was no exception; further, I have always liked the phrase attributed to Pasteur, "Chance favors the prepared mind." Curiosity may have killed the cat, but in this case it led yours truly to make some original contributions to the treasure of human knowledge just by keeping my eyes open and noticing subtle connections between various areas of research. Some of the contributions from my own research group are directly related to planarians.

One thing you must understand is that for most of us scientists, one of the highest things we can aspire to is for others to hear (or read) about our ideas and discoveries. An aspect of this is illustrated by the concept of the "gossip test," as described by one of science's greats, Dr. Francis Crick of DNA fame; we briefly mentioned this in chapter 3.[23] His reasoning was that one usually gossips (in the scientific sense) about things that one is truly passionate about; that passion fuels your determination and commitment and therefore increases your chances of success. A side effect (although not necessarily a bad one) of this passion is that you tend to go into the hallway at work and tell what you have just discovered in the lab quite literally to whomever you run into; this is usually preceded by a "Guess what I found?" This, of course, usually works best when talking to people who know you as opposed to random strangers in the hallway; otherwise, you may be perceived as odd, which again, does not necessarily have to be a bad thing. To be madly in love with science does not make you a mad scientist.

Anyway, since you, dear reader, and I are already acquainted, I'll feel free to tell you about my own planarian research story, which starts with a certain class of compounds that came from the sea.

FROM CORALS AND PLANTS TO PLANARIANS AND RATS

As we saw in previous chapters, nature produces a wide variety of chemicals used as weapons by living organisms. How much variety, you ask? A high estimate of the possible classes of small molecules in the context of pharmaceutical chemistry is about 1 followed by 60 zeroes.[24] There is no way that we can visualize the actual magnitude of such a number, let alone the "lab power" to test every possible compound's pharmacological activity. Please keep in mind that this does not mean that all those compounds exist in nature; this merely means that they are possible based on our current knowledge of organic chemistry. In this context, "small molecules" arbitrarily refers to compounds with a molecular weight of less than 800 Daltons or so. A "Dalton" is a chemical unit roughly equivalent to the weight of a hydrogen atom. In contrast, macromolecules, like antibodies, can have a molecular weight of about 150,000 Daltons. Most medications are considered small molecules. That said, bioinformatics tools are narrowing down the field, and while the number is still high, it facilitates the systematic study of drug candidates.[25]

Without any doubt, it is true that there are indeed a large number of classes of small molecules in nature. As we mentioned before, these chemicals oftentimes benefit the organism by being used for defense, predation, or both. Alternatively, some of these substances are also used to attract prospective mates, as in the case of pheromones. Some plants even use chemicals to lure other organisms, like insects, so they can serve as pollinating agents. Interestingly, in insect societies, like the ones represented by bees, ants, and termites, a variety of chemicals are used as a means of communication. As we saw before, we call the study of the use of chemicals by living organisms *chemical ecology*.[26] The development of this scientific discipline is providing us with the knowledge of a wide variety of compounds with interesting chemical structures that have proven useful in both fundamental research and practical applications, including but not limited to medical uses. The understanding of these structures is important, especially in the sense of how these substances interact with macromolecules like receptor targets. Please recall that "everything starts with binding" (see chapter 5)!

Marine organisms are a traditionally rich source of small molecules with interesting chemical properties, and in fact the story of how certain compounds came to be used in planarian research begins with a compound originally found in certain species of marine soft corals. This compound is called *lophotoxin*. There are several types of lophotoxinlike molecules, which share some distinct structural features (Figure 10.3).

Lophotoxin was originally isolated from soft corals of the genus *Lophogorgia* in 1981.[27] As the name implies, this molecule is used by corals as a defensive molecule (hence a toxin). Interestingly, lophotoxin and related molecules do not seem to be made by the coral itself; rather, they are synthesized by symbiotic microorganisms collectively called *zooxanthelae*.[28] This is not a unique or even especially uncommon arrangement in nature. Many species of life on earth develop associations with organisms of other species for mutual benefit. An important mission

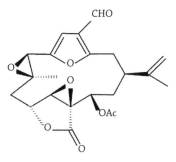

FIGURE 10.3. Lophotoxin.
Illustration by O. R. Pagán.

of scientific research is to discover and understand these relationships and look at how similar processes can be applied to the medical field.

Lophotoxin's mechanism of action involves the permanent inactivation of a protein closely involved in nerve and neuromuscular transmission in a wide variety of organisms, including humans. We'll take a brief detour to explore this important protein, called the *nicotinic acetylcholine receptor*.

Ion channels are proteins found in every single type of life forms; their main function is to regulate ion flow across cellular membranes. They are therefore important for communication between the cell and its external environment. We briefly saw an example of such channels when we mentioned local anesthetic effects. Local anesthesia functions by preventing the activation of neuronal voltage-gated sodium channels; this in turn prevents the transmission of nerve impulses, therefore blocking painful sensations. Yes, every time you go to the dentist, be grateful for ion channel blockers (see chapter 5 for an account on how the abused drug cocaine contributed to the development of local anesthesia).

One of the best-understood ion channels is the nicotinic acetylcholine receptor (nAChR). This is the receptor mainly responsible for muscle contraction in many animal species, ourselves included. When two molecules of the neurotransmitter acetylcholine (see chapter 3) bind to specific active sites (see chapter 5) on the receptors. This binding event initiates a series of physiological responses that lead to muscle contraction. The nAChR occupies a rather unique status among ion channels, as it was the first ion channel to be purified, allowing it to be studied at the biochemical level. nAChRs are defined by their specific pharmacology; they are the main biochemical target for nicotine; these receptors are therefore known as *nicotinic receptors* to us enthusiasts. It is very interesting that by far, nicotine directly targets nAChRs and virtually nothing else. Here "virtually" means that there are exceptions of course, which are beginning to be explored;

that said, the status of the nAChRs as the main targets for nicotine remains unchallenged.

One of the main reasons nicotinic receptors are so well known is that they can be isolated and purified in high amounts from the electric organs of certain fish, such as the marine ray (*Torpedo californica* and related species) and the freshwater electric eel (*Electrophorus electricus* and related species), common in certain parts of the Amazon. Developmentally, fish electric organs are related to muscle tissue, and they do exactly as the "electric" in their name implies; they produce electricity! These types of fish use their electric organs to stun their prey or to scare away predators. The electric rays are able to generate a voltage differential on the order of about 200 volts. For comparison, please remember that a normal electric plug (in the United States) carries about 110 volts. In ancient times, the Greeks used to step on electric rays to alleviate headaches and gout, anticipating modern physical therapy practices.[29] The Amazonian electric eel packs an even bigger punch; they can generate up to about 400 volts!

Electric rays and eels are able to generate such voltages and currents because their electric organs have two very important structural traits. The first one is a very high concentration of nicotinic acetylcholine receptors within their cells; actually, biochemists happily took advantage of their high concentrations in electric organs, where the receptor is about one thousand times more abundant than in vertebrate muscle.

The second important property of electric organs is that they are organized in very much the same way as a car battery. Since, as in any cells, there is a much higher concentration of sodium ions outside (remember that this is a simplified representation—see chapter 3), this concentration of electrically charged atoms generates these voltages within the electric organ. When the nicotinic receptor channels open, sodium rushes into the cells. When you have moving electrical charges, this is essentially electrical current, and due to the voltage caused by the batterylike arrangement of the cells in the electric organ, the generated currents are rather strong.

The study of nicotinic acetylcholine receptors is an important area of research. In many types of organisms, including humans, we already saw that there are versions of these receptors in skeletal muscle, where they control contraction. Interestingly, there are also versions of nicotinic receptors within nervous systems, including the brain itself, where they modulate phenomena like learning and memory, for example. Furthermore, in humans, many of these neuronal-type nAChRs most likely play a role in neurological conditions such as autism, Alzheimer's disease, and epilepsy, among others.

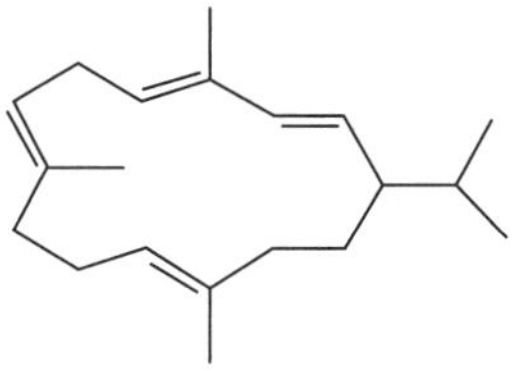

CEMBRENE

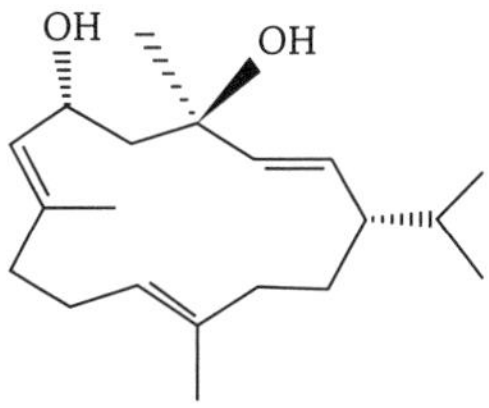

(4R)-CEMBRANETRIENEDIOL

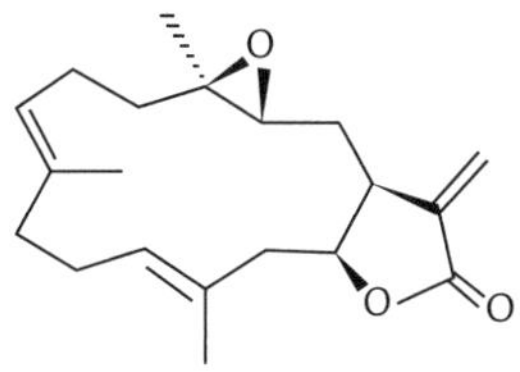

EUNIOLIDE

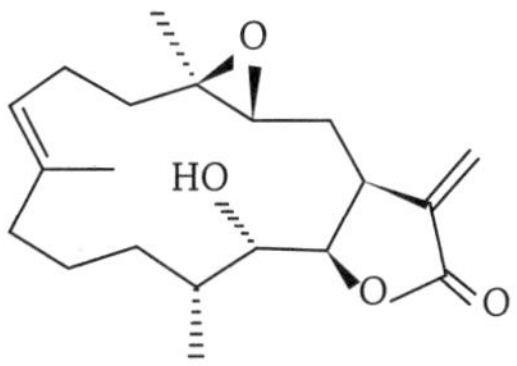

12,13-BISEPIEUPALMERIN

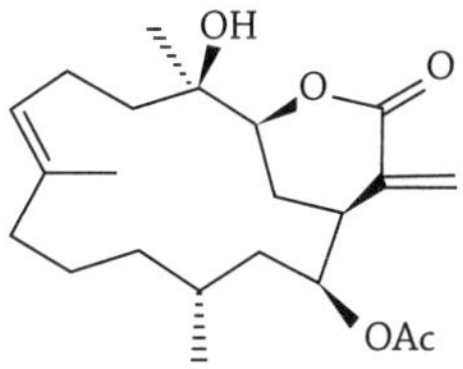

CRASSIN ACETATE

FIGURE 10.4. Representative cembranoidlike molecules. Please see Hann et al. (1998). Illustration by O. R. Pagán.

Well, back to lophotoxins. All lophotoxins share the same basic structure, which is built upon a so-called cembrene ring. This ring is composed of fourteen carbon atoms arranged in a specific way, to which a variety of different chemical groups are attached. These different chemical groups give the molecule their specific properties. In general, compounds that display a cembrene ring are called *cembranoids* (Figure 10.4).

Cembrenelike compounds were first described about fifty years ago, and they represent some of the most widespread types of compounds in the natural world. There are thousands of different naturally occurring and synthetic cembranoids reported in the scientific literature. This is significant in an evolutionary sense because if a particular chemical structure is abundant in nature, and especially if it is found in a wide variety of organisms, this usually means that it will likely confer a survival advantage to the organism. This is why, in my opinion, it is somewhat strange that this kind of compound is not more widely known outside the organic chemistry community, as cembranoids are some of the most interesting types of natural products.

Cembranoids have been isolated from a wide variety of organisms besides corals; they are produced in plants and insects. Some of the plants that produce cembranoidlike molecules include conifers and tobacco. Specifically in tobacco plants, there is evidence indicating that they serve as defensive compounds against pests.[30] These types of molecules are also found in insects such as ants and termites, where they show pheromonelike action and are a component of their chemical defense mechanism, respectively.[31,32] Cembranoids are not limited to plants or invertebrates. The male Chinese alligator (*Alligator sinensis*) and other crocodilian species secrete cembranoidlike molecules from its paracloacal glands.[33,34] Alligators and related scary critters probably use these secretions to mark their territory to either warn competitors against invading their space or to advertise their presence to potential mates. Crocodilians are the only vertebrate where cembranoids have ever been found so far, but life tends to conserve and reuse useful strategies. I would not be the least bit surprised if other vertebrates use these types of molecules as well. I think it is merely a matter of time before more examples are discovered.

In the meantime, by far, marine invertebrates are the richest cembranoid source in terms of quantity and variety. Most of these marine cembranoids are isolated from West Indian gorgonians. The latest comprehensive report in the scientific literature that I am aware of that provides a survey of this type of marine-based compound indicates that cembranoids make up about 25% of the total discovered metabolites in various coral species (this was published in 1995).[35] Surely this figure needs to be updated; nonetheless, I would like to say again that it stands to biological reason that if roughly a quarter of a given type of organism's metabolites share the same basic structure, this structure is bound to be important, no question about it. Nature tends to display good sense about these things; if something does not help a living being to survive, it tends to disappear.

Relatively few biological effects of the cembranoids have been described, but the most accepted interpretation is that as we saw, they represent defensive molecules. There is a large body of evidence that indicates that cembranoids are used as active chemical weapons by many types of coral species.

Biochemically, cembranoids from corals are known to act as antitumor compounds. Tobacco cembranoids, in addition to acting as defensive molecules, are also responsible for the characteristic aroma of tobacco products. Paradoxically, some tobacco cembranoids have even been found to possess antitumoral activity; this is strange, particularly since tobacco use is related to cancer. Also, some tobacco cembranoids have been reported to inhibit prostaglandin synthesis in some *in vitro* experiments. Prostaglandins are one of various types of compounds that are involved in a variety of physiological responses including inflammation; therefore, it is possible that cembranoids do show anti-inflammatory effects. Some cembranoids are even neuroprotective! If you are interested to know more about cembranoids in general, I refer you to a recent general review article.[36] At this time, we need to explore a little bit more the relationship between the cembranoids and nicotinic receptors.

Some cembranoids have been found to inhibit the physiological and behavioral effects of nicotine.[37] Interestingly, one of the cembranoids capable of preventing these nicotine effects is also produced in tobacco plants, the same source of nicotine! This is rather odd and presents an interesting puzzle. Think about it: why would a plant produce two types of substances that may act as physiological antagonists of each other? In other words, what nicotine does may be undone by a cembranoid and vice versa. This is kind of strange when you think about it; it is like giving a poison and its antidote at the same time. The physiological/evolutionary significance of the presence of these two substances in tobacco plants is a very interesting area of research that should be under intensive study.

I had the good fortune of being in the right place at the right time when the cembranoid–nicotine relationship began to be explored. Do you remember in the introduction when I said that I worked for about ten years at a medical school? That medical school is the Universidad Central del Caribe in my native Puerto Rico. I worked in the department of biochemistry under the supervision of three excellent scientists and educators, Drs. V. A. Eterovic, R. M. Hann, and P. A. Ferchmin. Based on the similarities of lophotoxin and related molecules with the cembranoids, they decided to explore the cembranoid–nicotinic receptor connection. I started working for them and then with them more than twenty years ago, and to this day, I have the pleasure and honor of counting them as friends and scientific collaborators. They gave me the opportunity to learn scientific research firsthand and believed in my capacities when I did not believe in them myself or even know that I really wanted to become a scientist. I am especially thankful for Dr. Hann's supervision when I was his technician for seven years and a fellow professor for another two. I learned to be an experimenter from him.

Eventually I wrote a master's thesis on the interaction of cembranoids with the nicotinic acetylcholine receptors and along the way became a coauthor of seven peer-reviewed papers and gained the experience and maturity that greatly facilitated my transition into a PhD program. Specifically, my master's thesis research was about cembranoids, nicotinic acetylcholine receptors, and anesthetic agents;[38] therefore, I became very familiar with these compounds. Also, another important detail is that I like chemical structures, and I particularly like cembranoidlike structures; this fact played an important role in this story.

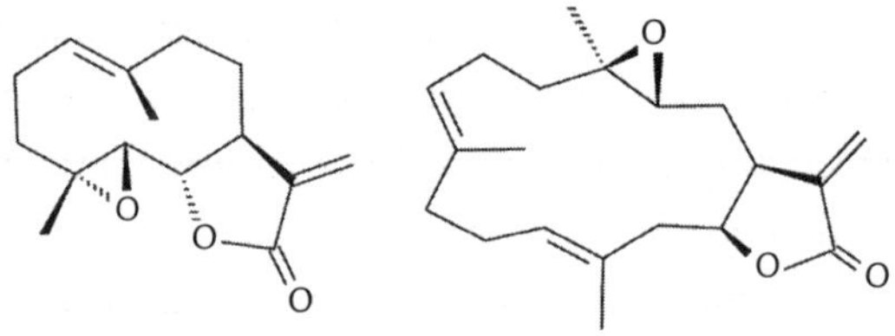

FIGURE 10.5. Parthenolide (at left); a representative cembranoid (euniolide, at right). Note the close structural similarities.
Illustrations by O. R. Pagán.

Where do planarians enter the storyline? Patience, patience...

Parthenolide (Figure 10.5) is a member of a chemical family collectively called the germacranolides. Germacranolides display herbicidal, fungicidal, and bactericidal activity in addition to showing antitumoral and anti-inflammatory properties *in vitro*. Parthenolide, in particular, is an inhibitor of serotonin type II receptors and interferes with microtubule formation. While mainly isolated from the feverfew plant *Tanacetum parthenium*, parthenolide is also found in other plant species. Feverfew extracts and teas are used in alternative medicine as treatment for migraines, arthritis, asthma, and other conditions. Parthenolide is only one of hundreds of compounds in the germacranolide class.

As interesting as these compounds sound, from the perspective of this story some of the main questions are: Why parthenolide? What is so special about it? Why did I decide to test it on planarians? There were several links that together tell this story (here comes the beer-pong–like connections scenario; I am immature this way).

- *Link 1:* I mentioned before that I like structures. I do not like them only because they are aesthetically pleasing; structures play fundamental roles in how molecules interact with other molecules (see chapter 5). One day I was browsing a catalog from a chemical company when the parthenolide structure caught my eye. It practically jumped off of the page because parthenolide displayed a very close structural similarity to certain cembranoids (Figure 10.5). I actually remember that the first thing I thought when I saw it was, "That looks like a mini cembranoid!"
- *Link 2:* Based on the work from my master's thesis, I knew that cembranoids and anesthetic agents like procaine bound in a mutually exclusive manner to the nicotinic acetylcholine receptor. Mutually-exclusive binding means that when one of them is bound to the target, the other is not. Further, this implies that these two compounds displace each other from

their respective binding sites; the simplest explanation, although not the only one, is that they are pharmacological competitors of each other.
- *Link 3:* Cocaine is the prototype of the local anesthetic superfamily of compounds that procaine is a part of, with neuronal sodium channels as the target (see chapter 5). However, procaine's behavioral effects are due to cocaine's interaction with neurotransmitter transporters, mainly the dopamine transporter as we saw before.
- *Link 4:* I tested parthenolide against cocaine on the nicotinic acetylcholine receptor and they did seem to compete with each other. These results are still unpublished.[39]
- *Link 5:* Again, parthenolide looks like a "mini" cembranoid (Figure 10.5).

In a nutshell, I reasoned that since certain cembranoids interacted with local anesthetics, cocaine was a local anesthetic, parthenolide interacted against cocaine in at least one type of receptor, and parthenolide looked like a cembranoid, why not test parthenolide against cocaine in an *in vivo* system, like planarians?

I love scientific connections.

Some preliminary experiments that I described in the last chapter of my PhD dissertation on the relationship between cocaine and parthenolide were promising, but not conclusive. I then graduated and started working at my current job to establish my own research group. Along the way, I kept thinking about planarians and I started working on them in my own laboratory from a pharmacological point of view.

Again, I really love the scientific process. During the course of trying to establish any effects of parthenolide by itself on planarians, I needed to dissolve parthenolide in the water the worms lived in. Now, parthenolide has the rather inconvenient tendency of not liking water so much, meaning that it does not dissolve that well in water. Such substances are termed *hydrophobic* (water fearing). To overcome that, I added a substance that helps dissolve such substances in water, called dimethylsulphoxide (let's call it DMSO for short). Why is this relevant? Read on...

When doing experiments, you have to be mindful to the best of your ability of any factor that can influence your results. Part of this is running something we call *control experiments*, which are basically the same experiments without the chemical that you really want to test. Therefore, as would any half-decent scientist, I ran some control experiments with just DMSO. I noticed something weird in my data.[40] With two (actually, my only two) research students,[41] I further explored these odd data and that led to my first scientific paper as an independent scientist.[42]

We then continued to work on the (still at that point) hypothetical relationship between cocaine and parthenolide in planarians. Our results eventually demonstrated that parthenolide prevents and rescues planarians from the effects of cocaine exposure, including motility decrease, seizure-like behaviors and withdrawal-like behaviors.[43,44] Thus, in planarians at least (and only in planarians), parthenolide and related compounds help decrease the effects of cocaine exposure. It cannot be emphasized enough that whether this translates to humans remains to be seen.

In a side story, as part of my interest in the parthenolide/cocaine connection, while still in graduate school (and with George's blessing), I started to collaborate with a colleague, Dr. Carlos Jiménez-Rivera, who leads a neurophysiology research group at the University of Puerto Rico, to try to test parthenolide against cocaine in rats. Basically, they gave cocaine to anesthetized rats and recorded the activity of a specific brain region, the ventral tegmental area (VTA). This brain region connects to many other areas of the brain and is very much related to addiction processes in vertebrates, including humans. When cocaine is given to a rat, there is a marked decrease in the neuronal activity of the VTA. Parthenolide reverses this reduction in activity; in other words, just like in planarians, parthenolide acts as a cocaine antagonist in rats, at least in neurophysiology experiments. These results were published.[45]

In addition to these intriguing results, we subsequently established that, at least in planarians, parthenolide specifically acts against cocaine-induced seizure-like behaviors and it does not interfere with such behaviors induced by amphetamines, nicotine, and so forth (see note 22).

On a related note, please remember that cembranoids act as nicotine antagonists. We tested a tobacco cembranoid against nicotine in planarians, and sure enough, this specific tobacco cembranoid reversed nicotine-induced behaviors in planarians,[46] just as it does *in vitro* and in rats.

So there we have it; in the cases of at least two substances abused by humans, planarians can serve as models for mammalian systems. This is part of the joy of understanding the intricately interwoven tapestry of nature.

PLANARIAN RESEARCH TRANSLATED TO VERTEBRATES: WHAT IT DOES MEAN, AND WHAT IT DOESN'T

All the results described in the last few paragraphs are intriguing from the point of view of fundamental research and show the practical potential of planarians in biomedical research. That said, the right thing to do is to state the appropriate "disclaimers."

Any nonhuman animal models that scientists may use in pharmacological research have one major drawback: they are not human. At the end of the proverbial day, clinical trials are not "strongly suggested" or mere "guidelines" but are required for legal approval and use of any medication. Are clinical trials perfect? No, they are not, not by a long shot. Part of the reason for this is that at this point in the development of clinical pharmacology, it is very difficult to predict the physiological (positive or negative) effects of a given compound in humans. This is especially difficult when dealing with totally new compounds. For example, if somebody comes up with a new cocainelike local anesthetic like lidocaine or procaine (novocaine), we have a lot of information about the specific mechanism of action of this class of compounds. Therefore, while clinical trials are still necessary, we are not going in blind with these types of compounds. On the other hand, new chemicals with lesser-understood biochemistry and pharmacology (like parthenolide, cembranoids, and related compounds) need to be looked at much more carefully. This is so since part of the reason that every drug displays side effects is because of their interaction with physiological targets that are not actually the intended targets to treat a given medical condition. These unknown effects have the potential to cause really bad situations. That said, by recognizing this, we've taken the first steps down this new road that shows a lot of promise.

Another very important consideration that needs to be kept in mind in light of the cembranoid–parthenolide–planarian–rat connection is that the cure for addiction is not around the corner, not based on planarian research or on any other research for that matter. There is a lot of work to be done in this area. Addiction is a very complex disease. We all know that addictive behaviors are not limited to the use and abuse of drugs. People can get addicted to a wide variety of things besides chemicals; gambling, certain foods, sex and even the internet are some of the most common addictions. Also, besides the well-established mechanisms that cause chemical dependence, there are many psychological aspects of addiction that are less understood. That being said, the physiological human correlates for addictive behavior are found in the brain, which brings me to my next point.

The planarian brain, as useful and interesting as it is, is much simpler than the human brain. This implies that despite this most welcomed success of the application of planarian research to mammals (think about the examples of parthenolide and the tobacco cembranoid that we discussed), any compounds proven useful to act against an abused drug in the planarian model may not act the same in vertebrate models, ourselves included. From the pure research perspective, the elucidation of the mechanism of the effects of a newly discovered compound in an animal model is incredibly interesting—no question about it—and I speak from direct experience. But when the objective is to develop useful medications intended for human administration, many other issues come into play, with safety as probably the most important issue, followed by a very close second, effectiveness.

That's the (kind of) bad news. Planarians may not be good human surrogates after all; we must not lose sight of this possibility.

The good news is that despite these possible limitations, which are shared by every other animal model that we care to examine, the use of planarians is proving itself as a valuable, relevant, and rather convenient animal model to try to explore neuropharmacology and, in fact, the neurosciences in general.

WHAT CAN THE PLANARIAN BRAIN TEACH US ABOUT OUR OWN?

In terms of fundamental neurobiology, as we have seen, the planarian brain shares multiple traits with more "advanced" nervous systems including ours. Please remember that in essence, the main point of a nervous system is to enhance the chances of survival of an organism through the behaviors that it generates. In this sense, and rather unexpectedly, in planarians we can see the rudiments of behaviors that were traditionally associated with more complex brains. These behaviors are ultimately controlled by the myriad chemicals that open/close/block/modulate the multiple kinds of receptors, transporters, enzymes, and many other proteins that make nervous systems work.

The best evidence that science has given us seems to indicate that brains arose for the first time as a means of dealing with an increasingly complex environment, which included the hostility from other organisms and competition for limited resources. Under these circumstances, having a brain was one way of endowing an organism with a distinct advantage over the efficient yet seemingly predictable and automatonlike actions of bacteria and other simpler organisms. With brains, likely for the first time, a living entity displayed an inkling of purposiveness as opposed to automaticity. For the first time, survival became something perhaps intentional rather than just incidental. Having a mechanism by which actions could be coordinated allowed complex organisms to develop more sophisticated survival strategies. Eventually, fully functional brains provided the means by which choices could be made, allowing organisms to exploit opportunities on one hand and flee danger on the other. Therefore, the appearance of even the most elementary organized neural activity provided a tremendous advantage to life and was therefore naturally selected and refined further over successive generations. Eventually, simple thoughts and even later the phenomenon of mind appeared on our planet. Here we have seen that in planarians, we see the first brain as a rudimentary, simple structure, one that nonetheless holds the promise of more complex neural organization, including our own.

It never ceases to amaze me how nature can surprise us, how much we can learn about ourselves just by carefully watching the fellow creatures that share the still poorly understood phenomenon of "life" with us. The history of the biomedical sciences regales us with plenty of examples of seemingly disparate biological entities that relate to each other in unforeseen ways. Each and every one of those examples can tell us something. Each of those examples can teach us invaluable lessons, but only if we learn how to listen to them. And our best hope to understand nature's lessons is called science.

I guess what I am trying to say can be expressed as a question: can the planarian brain help us understand our own?

Stranger things have happened.

Epilogue

IMAGINE A YOUNG girl in the present day, walking by a pond near Hobart, Tasmania, a heart-shaped island south of Australia; let's say that her name is Vanessa. She is exploring, looking for bugs, rocks, and other interesting things, hoping to collect all those little treasures that are of the utmost importance (and rightly so) when you are a curious child.

A small planarian wiggling in the water catches her eye. She is immediately fascinated by it. After all, what's not to like? This wormy is c-u-t-e, with its triangular head and its permanently crossed eyes. What she does not know is that this specific worm has a distinguished history.

This particular planarian came from the regeneration of an overlooked fragment of a worm that was eaten by a small fish about a month before. In turn, that worm that was (mostly) eaten by a fish came from yet another fragment that escaped the attention of a predatory insect. It is easy to imagine a long series of incredibly lucky worm fragments that were spared predation in an improbable but entirely possible chain of narrow escapes, stretching more than 170 years in the past. At that time, the "original" worm, which was hatched from an egg, was observed by a young man. A man that history records that kept the curiosity of a child until the day he died. This young man happened to be taking a stroll by the same pond as the little girl will do many years later. His name was Charles Robert Darwin.[1]

It is fun to daydream that essentially the same worm was observed by both the little girl and Charles Darwin, separated in time by almost two centuries. Their shared curiosity and wonder about nature represent some of the best traits in human nature; we believe that this curiosity is the hallmark of true intelligence. There are many strategies that organisms use to survive and eventually reproduce, which is the name of the game in evolution. In our case, we have survived as a species primarily because of our brain. Now, the brain of that little planarian tells it little more than to avoid light, to always swim against the current, to mate (sometimes), and to eat (at every possible opportunity). Their brains are so small that, as far as we know, it is not possible for them to wonder what they are or why they exist. However, that little planarian brain may well be one of our key tools to understanding ourselves; you see, part of the history of our own brain includes a nervous system very much like that tiny worm's. Because of our brain, we do wonder who we are, why we are here, and, further, where we are going. By studying those tiny, tiny brains from those tiny, tiny worms, we hope to be able to understand ourselves a little better.

Notes

CHAPTER 1

1. Snyder LJ, The Philosophical Breakfast Club, p. 3. This is a rather interesting book on that wonderful period in history when modern science was born.

2. Stewart I, The Mathematics of Life. This is a very readable, entertaining book that explores in a "user-friendly" manner a series of seemingly unrelated aspects of nature, especially biology, in the light of mathematics.

3. See Wigner E, The Unreasonable Effectiveness of Mathematics in the Natural Sciences.

4. See Cohen JE, Mathematics is Biology's Next Microscope, Only Better.

5. I have used the "wetness" concept as an example of an emergent property for a number of years now in my introductory biology lectures. I recently saw essentially the same example used in a book by Dr. Cristof Koch (Consciousness-Confessions of a Romantic Reductionist, p. 117). At first, I was kind of upset. No scientist wants to be scooped, but as far as I know he published the example first in book form, fair and square. I decided to take the attitude of "great minds think alike" and as I told this to my wife, I said "…at least it seems that I am thinking as a physicist." She answered, "Well, maybe *he* [her emphasis] is thinking like a biologist." I love that woman.

6. Kauffman S, At Home in the Universe.

7. Strogatz SH, Sync: How Order Emerges from Chaos in the Universe, Nature, and Daily Life.

8. Punset E, Excusas para no pensar, pp. 9–13.

9. Darwin CR, The Various Contrivances by Which Orchids Are Fertilised by Insects.

10. Ibid.

11. Ibid.

12. Gould SJ, Evolution in fact and theory. In: Hen's Teeth and Horse's Toes, pp. 253–62. Stephen Jay Gould was an accomplished evolutionary biologist and top-notch scientist all

around, but I think he reached more people than he'd ever imagined through his general science essays published in various venues, including the magazine *Natural History*, as well as in several popular science and technical books.

13. See Forterre P, Philippe H, The Last Universal Common Ancestor (LUCA), Simple or Complex?

CHAPTER 2

1. Maczulak A, Allies and Enemies: How the World Depends on Bacteria.
2. See Woese CR, A New Biology for a New Century.
3. Bonner JT, The Social Amoebae: The Biology of Cellular Slime Molds.
4. Conniff R, The Species Seekers: Heroes, Fools, and the Mad Pursuit of Life on Earth.
5. Yoon CK, Naming Nature: The Clash Between Instinct and Science.
6. Dunn R, Every Living Thing: Man's Obsessive Quest to Catalog Life, from Nanobacteria to New Monkeys.
7. It is important to clarify a very important point. There is this common misconception about the way we describe organisms that implies that some are "better" or "more advanced" than others. We must not confuse "more complex" with "advanced" or "better adapted." Evolution has no directionality. An organism can be simple in structural terms, yet perfectly adapted to its environment. Take parasites for example. Parasites are usually simpler organisms in the sense that they depend on other entities to survive. However, many of the most successful types of life forms are indeed parasites. Moreover, as we previously discussed (in chapter 1), organisms are "fit" or "adapted" invariably in the context of the environment in which they live.
8. http://cshprotocols.cshlp.org/site/emo/
9. http://www.thefreedictionary.com
10. Ibid.
11. Pagán OR, Synthetic Local Anesthetics as Alleviators of Cocaine Inhibition of the Human Dopamine Transporter, p. 1.
12. Wilson EO, The Future of Life., p. 120.
13. The Howard Hughes Medical Institute sponsors a series of Holiday Lectures where cutting-edge scientists explain their research in a layperson-friendly manner, understandable to a wide variety of audiences. Dr. Olivera's is found at http://www.hhmi.org/biointeractive/exploring-biodiversity-search-new-medicines.
14. Stott R, Darwin's Ghosts: The Secret History of Evolution, p. 29.

CHAPTER 3

1. Crick F, What Mad Pursuit: A Personal View of Scientific Discovery.
2. Rapport R, Nerve Endings: The Discovery of the Synapse. This is a delightful book. It reads like a novel and is very accurate, but most importantly, it is clearly written and quite entertaining. My favorite part is where it describes how Ramón y Cajal was "discovered" by the neurobiology scientific community. This part gave me goosebumps; it has the feeling of the dramatic turning point in a motion picture. Highly recommended.
3. Finger S, Minds Behind the Brain: A History of the Pioneers and Their Discoveries. A rather comprehensive view of the history of neuroscience.
4. The usual custom with Spanish surnames is to list the father's surname followed by the mother's surname (in English-speaking countries, the mother's surname is the maiden name).

Therefore, for Ramón y Cajal, "Ramón" is his father's surname and "Cajal" is his mother's. Following his own practice in his writings, I will refer to him as "Cajal" from now on.

5. Ramón y Cajal S, Recuerdos de mi vida. Tomo II - Historia de mi Labor Científica, pp. 490–91. The quote was translated from the Spanish original by Dr. Oné R. Pagán.

6. See DeFelipe J, Sesquicentenary of the Birthday of Santiago Ramón y Cajal, the Father of Modern Neuroscience.

7. Mazzarello P, Golgi: A Biography of the Founder of Modern Neuroscience.

8. Golgi C, Nobel lecture. This is the actual transcript of the lecture: http://www.nobelprize.org/nobel_prizes/medicine/laureates/1906/golgi-lecture.pdf

9. Ramón y Cajal S, Nobel lecture. This is the actual transcript of the lecture: http://www.nobelprize.org/nobel_prizes/medicine/laureates/1906/cajal-lecture.pdf

10. But see endnote 7 of chapter 2.

11. Bonner JT, The Social Amoebae: The Biology of Cellular Slime Molds.

12. Jones S, The Darwin Archipelago: The Naturalist's Career Beyond Origin of Species, p. 28. In fact, anyone should read ANY book by Dr. Jones.

13. Tompkins P, Bird C, The Secret Life of Plants.

14. See Brenner ED et al., Plant neurobiology: an integrated view of plant signaling.

15. See Rehm H, Gradmann D, Plant neurobiology, intelligent plants or stupid studies.

16. Chamovitz D, What a Plant Knows: A Field Guide to the Senses. A brief yet informative book on the topic.

CHAPTER 4

1. Please see footnote 7, chapter 2.

2. Swanson LW, Brain Architecture: Understanding the Basic Plan.

3. See Swanson LW, What is the brain?.

4. This is an important question that is also the title of a paper particularly relevant to this discussion: Sarnat HB and Netsky MG, When does a ganglion become a brain?.

5. See Netsky MG, What is a brain, and who said so?.

6. See Gross CG, Aristotle on the brain.

7. Finger S, Minds Behind the Brain: A History of the Pioneers and Their Discoveries

8. Ibid.

9. Marshall LH, Magoun HW, Discoveries in the Human Brain: Neuroscience Prehistory, Brain Structure, and Function.

10. Zimmer C, Soul Made Flesh: The Discovery of the Brain–And How It Changed the World.

11. Azevedo FA et al., Equal numbers of neuronal and nonneuronal cells make the human brain an isometrically scaled-up primate brain. The actual figure for the calculated number of brain cells in this paper above is 86.1 ± 8.1. This is based on a sample of four male human brains. Note that based on this, the upper limit seems to be about 94 billion neurons, not so different from 100 billion. Moreover, it can go as low as 78 billion, but no one mentions this low figure. The fact that this is not explained in popular science articles is to me an example of bite-size science, one of my pet peeves. This is a story for another time though. Anyway, this was not even the point of the paper and the authors never stated that it was, but the "86 billion" number stuck. More experiments are needed to reliably ascertain this figure. Therefore, for the rest of the book, I'll stick with a nice round number (100 billion). In a recent development, one of the authors of the Azevedo paper, Dr. Roberto Lent, told me via email that a new paper is under way extending their results with a larger data sample. This is very exciting! However, at the time of this writing, the newest paper has not yet been published.

12. Seung S, Connectome. This is a rather optimistic, very well-written, and entertaining introduction to the connectome concept.

13. Hagmann P, From Diffusion MRI to Brain Connectomics.

14. See Sporns O et al., The human connectome: a structural description of the human brain.

15. Fields RD, The Other Brain: From Dementia to Schizophrenia, How New Discoveries About the Brain Are Revolutionizing Medicine and Science.

16. This is based on a quote from Dr. Stephen Smith of Stanford University: "One synapse, by itself, is more like a microprocessor—with both memory-storage and information-processing elements—than a mere on/off switch. In fact, one synapse may contain on the order of 1,000 molecular-scale switches. A single human brain has more switches than all the computers and routers and Internet connections on Earth." Downloaded from http://med.stanford.edu/ism/2010/november/neuron-imaging.html.

17. See DeFelipe J, From the connectome to the synaptome: an epic love story.

18. Alivisatos AP et al. The brain activity map project and the challenges of functional connectomics. Neuron 74:970–74. This is an excellent review paper that summarizes one of the approaches that is being used currently to functionally map brain processes.

CHAPTER 5

1. See Rounak S, Zoopharmacognosy (animal self-medication): a review.

2. See Huffmann MA, Current evidence for self-medication in primates: a multidisciplinary perspective.

3. Eisner T and Meinwald J, Editors, Chemical Ecology: The Chemistry of Biotic Interaction

4. See Cobo B, Historia del Nuevo Mundo. Manuscrito en Lima, Perú, 1653, libro 5°, capítulo XXIX. Modern edition: Bernabé Cobo. Historia del Nuevo Mundo. Sevilla. Publicado por la Sociedad de Bibliófilos Andaluces. Con notas de Marcos Jiménez de la Espada. Impreso por E. Rasco, 1890, Vol I, Book 5th, Chapter XXIX, pp. 473–77. Cited in Calatayud J, González A (2003) History of the development and evolution of local anesthesia since the coca leaf.

5. Streatfeild D, Cocaine: An Unauthorized Biography.

6. See Hertting G, Axelrod J, Fate of tritiated noradrenaline at the sympathetic nerve endings. Dr. Julius Axelrod is one of my favorite scientists ever. He started his career as a technician, got his PhD in his forties and went on to win a Nobel Prize. Along the way, he mentored quite a cadre of scientists that lists many prominent figures in neuropharmacology. A fascinating scientific life!

7. Elrich PR, Raven PH, Butterflies and plants.

8. See Sullivan RJ, Hagen EH, Psychotropic substance-seeking: evolutionary pathology or adaptation?

9. See Keesey J, How electric fish became sources of acetylcholine receptor.

CHAPTER 6

1. See Rieger RM, 100 years of research on 'Turbellaria'.

2. Internet sources report this worm to be about 600 mm; it is unclear where this figure came from. However, I was able to obtain a copy of the original paper thanks to Dr. Masaharu Kawakatsu: See Porfireva NA (1969) Endemichnaia Baikal'skaia Planariia *Rimacephalus arecepta* sp. n. (Tricladida, Paludicola) (An endemic Baikal planarian *Rimacephalus arecepta* sp. n. [Tricladida, Paludicola]). This publication shows a picture of a specimen 90 mm long.

3. Personal communication from Dr. Emili Saló and Dr. Jaume Baguñà, Department of Genetics, University of Barcelona, Spain. This department has a long tradition of quality flatworm-related research.

4. See Kawakatsu M et al., Bipalium nobile sp.nov. (Turbellaria,Tricladida,Terricola), a New Land Planarian from Tokyo.

5. See Davies L et al., Nerve repair and behavioral recovery following brain transplantation in *Notoplana acticola*, a polyclad flatworm.

6. See Koopowitz H, Holman M, Neuronal repair and recovery of function in the polyclad flatworm. Dr. Koopowitz and collaborators published several interesting papers on polyclad brains—worth looking at!

7. See Saito Y et al., Mediolateral intercalation in planarians revealed by grafting experiments.

8. See Stephan-Dubois F, Gilgenkrantz F, Transplantation and regeneration in the planarian, *Dendrocoelum lacteum*.

9. Miller JA, Studies on heteroplastic transplantation in triclads I. Cephalic grafts between *Euplanaria dorotocephala* and *E. tigrina*.

10. See Egger B et al., To be or not to be a flatworm: the acoel controversy.

11. See Maxmen A, A can of worms.

12. See Baguñà J et al., Back in time: a new systematic proposal for the Bilateria.

13. Ibid.

14. See Northcutt RG, Evolution of centralized nervous systems: two schools of evolutionary thought.

15. See Erwin DH, Davidson EH, The last common bilaterian ancestor.

16. Ibid.

17. See Poinar G, A rhabdocoel turbellarian (*Platyhelminthes, Typhloplanoida*) in Baltic amber with a review of fossil and sub-fossil platyhelminths.

18. See Allison CW, Primitive fossil flatworm from Alaska: new evidence bearing on ancestry of the Metazoa.

19. See Cloud P et al, Traces of animal life from 620-million-year-old rocks in North Carolina.

20. See Pierce WD, Silicified turbellaria from calico mountains nodules.

21. See Dentzien-Dias PC et al., Tapeworm eggs in a 270 million-year-old shark coprolite.

22. See Egger B et al., Free-living flatworms under the knife: past and present.

23. Jenkins MM, Aspects of planarian biology and behavior, p. 116–43. This is an invaluable book including many examples of planarian research, in particular about their behavior.

24. http://www.pbs.org/kcet/shapeoflife/episodes/hunter.html

25. Brøndsted HV, Planarian Regeneration (International Series of Monographs in Pure and Applied Biology. Division: Zoology).

26. Hyman LH andHutchinson, EG, Libbie Hyman 1888–1969. A biographical memoir.

27. Winston JE, Libbie Henrietta Hyman: life and contributions.

28. Komai T, Kawakatsu M, In memoriam - Dr. Libbie H. Hyman.

29. Kawakatsu M, North American triclad Turbellaria, 17: freshwater planarians from Lake Tahoe.

30. Kawakatsu M [Collection and observation of plants in the vicinity of Sonobe-chô, Kyôto Prefecture]. (in Japanese).

31. Okugawa KI, An experimental study of sexual induction in the asexual form of Japanese fresh-water planarian, *Dugesia gonocephala* (Dugès).

32. Kawakatsu M, On the ecology and distribution of freshwater planarians in the Japanese Islands, with special references to their vertical distribution.

33. Sluys R, Kawakatsu M, Bleeker J The Kawakatsu Collection incorporated within the collection of the Zoological Museum Amsterdam.

34. I wish to express my gratitude to Dr. Vida Kenk, an invertebrate biologist in her own right, who generously shared her family history, documents, and other materials with me, both through email and during a delightful telephone conversation. My connection with Dr. Roman Kenk is fourfold. First is the obvious planarian connection. Also, he worked at the University of Puerto Rico, where I earned my bachelor's and master's degrees. During his time in Puerto Rico he described a Puerto Rican planarian species (*Dugesia antillana*, Figure 3.5). Finally, he married someone with the Pagán surname! I love connections.

CHAPTER 7

1. Holmes R, The Age of Wonder: The Romantic Generation and the Discovery of the Beauty and Terror of Science.

2. Snyder LJ, The Philosophical Breakfast Club: Four Remarkable Friends Who Transformed Science and Changed the World, p. 3.

3. Francis RC, Epigenetics: The Ultimate Mystery of Inheritance.

4. See Moore R, The "rediscovery" of Mendel's work.

5. Ibid.

6. See Pigliucci M, Do we need an extended evolutionary synthesis?.

7. Randolph H, Observations and experiments on regeneration in planarians.

8. Mitman G, Fausto-Sterling A, Whatever happened to planaria? C.M. Child and the physiology of inheritance. In: Clarke AE, Fujimura JH, Editors. The Right Tools for the Job, pp. 172–97.

9. See Elliot SA, Sánchez Alvarado A, The history and enduring contributions of planarians to the study of animal regeneration.

10. Wharton DA, Life at the Limits: Organisms in Extreme Environments.

11. See Egger B, Regeneration: rewarding, but potentially risky.

12. Bely AE, Nyberg KG, Evolution of animal regeneration: re-emergence of a field.

13. Mather JA et al., Octopus: The Ocean's Intelligent Invertebrate, p. 100.

14. See Glauber KM et al., Hydra.

15. See Galliot B, Hydra, a fruitful model system for 270 years.

16. See Shimizu H et al., Minimum tissue size required for hydra regeneration.

17. See Holstein TW et al., Cnidarians: an evolutionarily conserved model system for regeneration?.

18. See Agata K, Inoue T, Survey of the differences between regenerative and non-regenerative animals. In fact, the discovery of the apparent molecular pathway responsible for head regeneration in planarians was recently reported in three recent published in Nature (Liu SY et al., Reactivating head regrowth in a regeneration-deficient planarian species; Sikes JM and Newmark PA, Restoration of anterior regeneration in a planarian with limited regenerative ability; and Umesono Y et al., The molecular logic for planarian regeneration along the anterior-posterior axis. These papers were profiled in the same issue of Nature by Simon A, Regenerative biology: On with their heads).

19. See note 11.

20. See Morgan TH, Experimental studies of the regeneration of Planaria maculata.

21. See Montgomery JR, Coward SJ, On the minimal size of a planarian capable of regeneration.

22. Ibid.

23. It is so good to have smart friends! My thoughts on the matter of the minimal size/cell number for planarian regeneration were fine-tuned by a great email conversation with three experts on the area, Dr. Emili Saló (University of Barcelona, Spain, who incidentally brought to my attention that dissociated hydra will reorganize into whole animals), Dr. Mike Levin (Tufts University, Boston, MA, USA) and Dr. Néstor Oviedo (University of California at Merced, USA).

24. See Egger B, The regeneration capacity of the flatworm *Macrostomum lignano*—on repeated regeneration, rejuvenation, and the minimal size needed for regeneration; and Ladurner P et al., A new model organism among the lower Bilateria and the use of digital microscopy in taxonomy of meiobenthic Platyhelminthes: *Macrostomum lignano*, n. sp. (Rhabditophora, Macrostomorpha).

25. Ibid.

26. See Agata K et al., Two different evolutionary origins of stem cell systems and their molecular basis.

27. See Wagner DE et al., Clonogenic neoblasts are pluripotent adult stem cells that underlie planarian regeneration.

28. See Benazzi M et al., Further contribution to the taxonomy of the *Dugesia lugubris-polychroa* group with description of *Dugesia mediterranea* n. sp. (Tricladida, Paludicola).

29. Cebrià F et al., Myosin heavy chain gene in *Dugesia (G.) tigrina*: a tool for studying muscle regeneration in planarians.

30. Vispo M et al., Regionalisation along the anteroposterior axis of the freshwater planarian *Dugesia(Girardia)tigrina* by TCEN49 protein.

31. Powell K, Masters of regeneration, pp. 28–48. The funny thing about this story is that Newmark and Sánchez Alvarado brought the planarians they caught in Barcelona back to the United States inside ice coolers wrapped in duct tape *in the airplane's passenger cabin*. They report that they got suspicious looks but that they were able to go through. This would never have happened for obvious reasons just a few years later.

32. http://smedgd.neuro.utah.edu/

33. See Reddien PW et al., Gene nomenclature guidelines for the planarian *Schmidtea mediterranea*.

34. See Baguñà J, The planarian neoblast: the rambling history of its origin and some current black boxes.

35. See Saló E, Baguñà J, Regeneration and pattern formation in planarians. I. The pattern of mitosis in anterior and posterior regeneration in *Dugesia (Girardia) tigrina*, and a new proposal for blastema formation.

36. See Agata K, Watanabe K, Molecular and cellular aspects of planarian regeneration.

37. See Saló E et al., Planarian regeneration: achievements and future directions after 20 years of research.

38. Brøndsted HV, Planarian Regeneration.

39. Aboobaker AA, Kao D, A lack of commitment for over 500 million years: conserved animal stem cell pluripotency.

40. Tan TC et al., Telomere maintenance and telomerase activity are differentially regulated in asexual and sexual worms.

41. Aboobaker AA, Planarian stem cells: a simple paradigm for regeneration.

CHAPTER 8

1. A tongue-in-cheek satirical quote from McConnell JV, Schutjer M, Science, Sex, and Sacred Cows: Spoofs on Science from the Worm Runner's Digest, p. 108.

2. Actually, they can perceive light, but they cannot "see" the way vertebrates do (see chapters 6 and 7).

3. Larsen L, In All Their Animal Brilliance: Poems. Used with permission.

4. An early version of this section was originally published in my *Baldscientist* blog: http://baldscientist.wordpress.com/2011/11/02/planarians-and-the-new-battlestar-galactica/

5. Di Justo P, Grazier K. The Science of Battlestar Galactica.

6. http://www.guardian.co.uk/books/2012/dec/04/paul-krugman-asimov-economics. I wish to thank Dr. Krugman for his help and for taking the time to point me in the right direction to find the relevant references.

7. Dalyell JG, (Observations on Some Interesting Phenomena in Animal Physiology, Exhibited by Several Species of Planariae: Illustrated by Coloured Figures of Living Animals.

8. Brøndsted HV, Planarian Regeneration (International Series of Monographs in Pure and Applied Biology. Division: Zoology).

9. The Bible. Ecclesiastes 1:9; HCSB Digital Text Edition.

10. This is actually a "triple connection," since as we saw in the previous chapter, the journal *Physiological Zoology* was founded by C. M. Child in no small part because of his rivalry with T. H. Morgan. Please remember that both Child and Morgan were arguably the two main characters in the early history of the scientific exploration of regeneration using planarians (see chapter 6).

11. See Silber RH, Hamburger V, The production of duplicitas cruciata and multiple heads by regeneration in *Euplanaria tigrina*, pp.11, 13–15.

12. *The Big Bang Theory* episode 17: "The Monster Isolation"; © 2013 CBS.

13. See Lenicque PM et al., Effects of some components of *Cannabis sativa* on the regenerating planarian worm *Dugesia tigrina*.

14. See Buttarelli FR et al., Cannabinoid-induced stimulation of motor activity in planaria through an opioid receptor-mediated mechanism.

15. See Rawls SM et al., Agmatine: identification and inhibition of methamphetamine, kappa opioid, and cannabinoid withdrawal in planarians.

16. *Dr. Who Fourth Series* episode 13: "Journey's End"; © 2013 BBC.

CHAPTER 9

1. See Li YH, Tian X, Quorum sensing and bacterial social interactions in biofilms.

2. See Fuqua WC et al., Census and consensus in bacterial ecosystems: the LuxR-LuxI family of quorum-sensing transcriptional regulators.

3. See Raffa RB et al., Bacterial communication ("quorum sensing") via ligands and receptors: a novel pharmacologic target for the design of antibiotic drugs.

4. Bonner JT, The Social Amoebae: The Biology of Cellular Slime Molds.

5. See Saxe CL, Learning from the slime mold: Dictyostelium and human disease.

6. See Nakagaki T et al., Smart behavior of true slime mold in a labyrinth.

7. See Nakagaki T Maze-solving by an amoeboid organism.

8. See Saigusa T et al., Amoebae anticipate periodic events.

9. See Adams ED et al., Freshwater sponges have functional, sealing epithelia with high transepithelial resistance and negative transepithelial potential.

10. See Elliott GR, Leys SP, Evidence for glutamate, GABA and NO in coordinating behaviour in the sponge, *Ephydatia muelleri* (Demospongiae, Spongillidae).

11. Colansanti A, Venturini G, Nitric oxide in lower invertebrates. In: Raffa RB, Rawls SM, Editors. Planaria: A Model for Drug Action and Abuse, pp. 31–39.

12. Lentz TL, Primitive Nervous Systems, p. 19. This is a brief yet seminal work on the exploration of the structure, function, and evolution of early nervous systems.

13. Matheson T, Invertebrate nervous systems. Also see Lentz TL, Primitive Nervous Systems.

14. See Arendt D et al., The evolution of nervous system centralization.

15. *Elysia chlorotica*. http://eol.org/pages/450768/overview

16. See http://www.nature.com/news/photosynthesis-like-process-found-in-insects-1.11214 and Valmalette JC et al., Light- induced electron transfer and ATP synthesis in a carotene synthesizing insect.

17. See http://www.nature.com/news/2010/100730/full/news.2010.384.html and Kerney R et al., Intracellular invasion of green algae in a salamander host.

18. See Moroz LL, On the independent origins of complex brains and neurons.

19. See Northcutt RG, Evolution of centralized nervous systems: two schools of evolutionary thought.

20. http://www.pbs.org/kcet/shapeoflife/episodes/hunter.html. This is a rather entertaining documentary from PBS/National Geographic as part of its series "The Shape of Life."

21. I do not know about you, but oftentimes I don't even know what *I'm* thinking, let alone what the little worms in my laboratory are thinking.

22. See Sarnat HB, Netsky MG, The brain of the planarian as the ancestor of the human brain. This is the paper that triggered my interest in using planarians in neurobiology.

23. I feel very strongly about this. In fact, one of my latest scientific papers includes the following sentence: "Given the power of modern molecular techniques, it is easy to forget that the ultimate objective of physiological discoveries is to find their possible significance within the context of the whole organism, particularly its behavior." Quoted from Pagán OR et al., Planarians require an intact brain to behaviorally react to cocaine, but not to react to nicotine.

24. See Bailly X et al., The urbilaterian brain revisited: novel insights into old questions from new flatworm clades; and Reichert H, Simeone A, Developmental genetic evidence for a monophyletic origin of the bilaterian brain. These are especially clear and thorough review articles on the current state of the art of the exploration of the early brain.

25. See Agata K et al., Structure of the planarian central nervous system (CNS) revealed by neuronal cell markers.

26. See Murakoshi H, Yasuda R, Postsynaptic signaling during plasticity of dendritic spines.

27. See Nimchinsky EA et al., Structure and function of dendritic spines.

28. See Lentz TL (1968); Primitive nervous systems. Also see Nishimura K et al, Brain and neural networks. In: Raffa RB, Rawls SM, Editors. Planaria: A Model for Drug Action and Abuse, pp. 4–12.

29. See Bailly X et al., The urbilaterian brain revisited: novel insights into old questions from new flatworm clades; Reuter M, Gustafsson MK, The flatworm nervous system: pattern and phylogeny; Richter S et al., Invertebrate neurophylogeny: suggested terms and definitions for a neuroanatomical glossary.

30. See Aoki R et al., Recording and spectrum analysis of the planarian electroencephalogram.

31. As a science enthusiast, you may have heard of Stephen Hawking, a physicist and cosmologist who is one of the finest minds that humanity currently has. He has ALS; it has not stopped his mind at all.

32. See Cebrià F et al., FGFR-related gene *nou-darake* restricts brain tissues to the head region of planarians. *Nature* is one of the two "top" scientific journals that publish original research. The other one is *Science*. Most scientists (including yours truly) dream of having a paper in either one of those journals.

33. See Almuedo-Castillo M et al., Wnt signaling in planarians: new answers to old questions; Gurley KA, et al., Beta-catenin defines head versus tail identity during planarian regeneration and homeostasis; Niehrs C, On growth and form: a Cartesian coordinate system of Wnt and BMP signaling specifies bilaterian body axes; Petersen CP, Reddien PW, Smed-betacatenin-1 is required for anteroposterior blastema polarity in planarian regeneration.

34. There are plenty of resources available to explore this further. It will depend on how deep you want to go. Two useful general resources are http://www.exploratorium.edu/imaging-station/research/planaria/story_planaria.pdf and http://www.scholarpedia.org/article/Planaria_nervous_system. More specialized resources include Raffa RB, Rawls SM, Editors, Planaria: A Model for Drug Action and Abuse; Bailly X et al., The urbilaterian brain revisited: novel insights into old questions from new flatworm clades; Cebrià F, Regenerating the central nervous system: how easy for planarians!; Gentile L et al., The planarian flatworm: an in vivo model for stem cell biology and nervous system regeneration; Reuter M, Gustafsson MK, The flatworm nervous system: pattern and phylogeny; Richter S et al., Invertebrate neurophylogeny: suggested terms and definitions for a neuroanatomical glossary; Saló E et al., Planarian regeneration: achievements and future directions after 20 years of research.; Umesono Y et al., Evolution and regeneration of the planarian central nervous system.

35. Mather JA et al., Octopus: The Ocean's Intelligent Invertebrate.

36. Williams W. Kraken: The Curious, Exciting, and Slightly Disturbing Science of Squid. Very readable, entertaining book. I was pleasantly surprised that it included a fair amount of neurobiology.

37. Denes AS et al., Molecular architecture of annelid nerve cord supports common origin of nervous system centralization in bilateria.

38. Bainbridge D, Beyond the Zonules of Zinn: A Fantastic Journey Through Your Brain, Chapter 6. This is an especially delightful layperson account of this story in light of the history of science.

39. Best JB, Protopsychology.

40. See Walter HE, The reactions of planaria to light. This scientific paper is about 135 pages long!

41. Ichikawa A, Kawakatsu M, A new freshwater planarian *Dugesia japonica*, commonly but erroneously known as *Dugesia gonocephala* (Dugès).

42. See van Oye P, Over het geheugen bij de Platwormen en andere Biologische waarnemingen bij deze dieren.

43. Wells PH, Training planarians in a van Oye maze. In: Corning WC, Ratner SC, Editors. The Chemistry of Learning: Invertebrate Research. pp. 251–54.

44. Rilling M, The mystery of the vanished citations: James McConnell's forgotten 1960s quest for planarian learning, a biochemical engram, and celebrity. This is a well-written and balanced account of the love–hate relationship between McConnell and science. In here it says that McConnell's advisor, Dr. M. E. Bitterman suggested the use of planarians to McConnell; oddly, in a personal account (McConnell JV, Cannibalism and memory in flatworms), he credited his lab mate, Richard Thompson, as the one who introduced him to planarians. Perhaps both versions are (kind of) correct to some extent?

45. This act by McConnell reminded me of Galileo's well-known fact that he wrote one of his popular science books (Dialogue Concerning the Two Chief World Systems) in the form of a conversation between three characters, Salviati, an intellectual; Sagredo, an intelligent layperson; and a third character called Simplicio (as in simpleton). To make a long story short, some of the words that Galileo put in the mouth of Simplicio came directly from Pope Urban VIII. Let's just say that it was not a good idea to offend the Holy Inquisition's boss.

46. See Rilling M, The mystery of the vanished citations: James McConnell's forgotten 1960s quest for planarian learning, a biochemical engram, and celebrity.

47. See Thompson R, McConnell J, Classical conditioning in the planarian, *Dugesia dorotocephala*.

48. See Corning WC, John ER, Effect of ribonuclease on retention of conditioned response in regenerated planarians.

49. See Bennett EL, Calvin M, Failure to train planarians reliably.

50. See McConnell JV, Comparative physiology: learning in invertebrates.

51. McConnell JV, Cannibalism and memory in flatworms. This is essentially a magazine article. If you want to know more, I suggest the references that follow.

52. Shomrat T, Levin M, An automated training paradigm reveals long-term memory in planaria and its persistence through head regeneration.

53. Blackiston D et al., A second-generation device for automated training and quantitative behavior analyses of molecularly-tractable model organisms.

54. Corning WC, Ratner SC, Editors, The Chemistry of Learning: Invertebrate Research.

55. Boese A, Elephants on Acid: And Other Bizarre Experiments. Chapter 3; Collins HM, Pinch T, The Golem: What You Should Know About Science, Chapter 1.

CHAPTER 10

1. Wilson EO, The Future of Life, p. 120. One of my favorite books by one of my favorite scientists. This book makes a series of very important points, including that even if we do not realize it, whether we like it or not, we are all intimately related to every other form of life on earth. The book further argues for the conservation and protection of our environment. You'd expect that from a biologist after all, but Dr. Wilson goes beyond the mere "save the earth" frame of mind. He presents in a very clear way many arguments that explain why it is a good idea for us to do so. He expresses the usual arguments such as that we are stewards of the planet and similar thoughts, but he also makes the case that it makes economic sense as well.

2. Pagán OR, Synthetic Local Anesthetics as Alleviators of Cocaine Inhibition of the Human Dopamine Transporter, p. 1.

3. See Needleman HL, Tolerance and dependence in the planarian after continuous exposure to morphine.

4. I like Dr. Needleman not so much because of his planarian work (it seems that he only published one paper on the subject) but because he was probably the main character that brought to light and offered solutions to one of the most high-profile public health issues, the toxicity of lead exposure, especially in children. As a direct consequence of his efforts, legislation was drafted to eliminate lead from gasoline and paint products, with undeniable positive benefits for public health. This is an interesting story documented in various venues. Two very useful published sources are Rosner D and Markowitz G, Standing up to the lead industry: an interview with Herbert Needleman; and Needleman H, The case of Deborah Rice: who is the Environmental Protection Agency protecting?

5. See Best JB, Morita M, Toxicology of planarians.

6. See Pedersen KJ, Morphogenetic activities during planarian regeneration as influenced by triethylene melamine.

7. See Rawls SM et al., Nicotine behavioral pharmacology: clues from planarians.

8. See Pagán OR et al, A cembranoid from tobacco prevents the expression of nicotine-induced withdrawal behavior in planarian worms.

9. See Raffa RB, Desai P, Description and quantification of cocaine withdrawal signs in planaria.

10. Anonymous, Microscopy.

11. See Heath H, A new turbellarian from Hawaii.

12. See Palladini G et al., A pharmacological study of cocaine activity in planaria.

13. See Margotta V et al., Effects of cocaine treatment on the nervous system of planaria (Dugesia gonocephala s. l.).

14. See Carolei A et al., Proposal of a new model with dopaminergic-cholinergic interactions for neuropharmacological investigations.

15. Ichikawa A, Kawakatsu M, A new freshwater planarian *Dugesia japonica*, commonly but erroneously known as *Dugesia gonocephala* (Dugès).

16. See Buttarelli FR et al., Neuropharmacology and behavior in planarians: translations to mammals.

17. Raffa RB, Valdez JM, Cocaine withdrawal in Planaria. Personally, I credit this very paper for my own interest on using planarians as a research subject in pharmacology and therefore for my own research program.

18. Professor Hess is currently retired from a long and productive career that spanned some fifty-five years at Cornell where he is currently a Professor Emeritus. He has published hundreds of papers, and among his many honors, he is a member of the National Academy of Sciences, the "who's who" of science in the United States, roughly equivalent to the Royal Society in Britain. Cornell has a rather convenient arrangement for graduate students. In addition to the academic departments, the areas of study are organized into "fields," which essentially are concentrations or majors. For example, my own field was pharmacology, whose offices were located at the Department of Molecular Medicine, but I chose to work under a researcher at the Department of Molecular Biology and Genetics. I also was "minoring" in neurobiology as a subfield. The net effect was that I was able to use the resources and equipment of several departments without any hassle whatsoever. Quite nifty!

19. It was very hard for me to be on a first-name basis with my professors. It was particularly intimidating with George. About two weeks after I started working in his lab, I was in his office discussing research. When I was about to leave, I responded to something he said with a "Yes, sir." He then said, "Drop the 'Sir'; call me George. Now, go do some work." That was it. From then on, he was simply "George."

20. Please see note 2.

21. See Torres GE et al., Plasma membrane monoamine transporters: structure, regulation and function.

22. Pagán OR et al., Planarians in pharmacology: parthenolide is a specific behavioral antagonist of cocaine in the planarian *Girardia tigrina*. This was a paper that we were invited to submit as part of the *International Journal of Developmental Biology* special issue on planarian biology.

23. Crick F, What Mad Pursuit: A Personal View of Scientific Discovery, Chapter 2.

24. Stockwell BR, The Quest for the Cure: The Science and Stories Behind the Next Generation of Medicines.

25. See Hann MM, Keserü GM, Finding the sweet spot: the role of nature and nurture in medicinal chemistry.

26. Eisner T, Meinwald J, Editors, Chemical Ecology: The Chemistry of Biotic Interaction.

27. See Fenical W et al., Lophotoxin: a novel neuromuscular toxin from Pacific sea whips of the genus *Lophogorgia*.

28. See Ciereszco LS, Sterol and diterpenoid production by Zooxanthellae in coral reefs: a review.

29. See Keesey J, How electric fish became sources of acetylcholine receptor.

30. See Wang E et al., Transgenic *Nicotiana tabacum l.* with enhanced trichome exudate cembratrieneols has reduced aphid infestation in the field.

31. See Prestwich GD, Interspecific variation of diterpene composition of *Cubitermes* soldier defense secretion.

32. See Wiemer DF et al., Cembrene A and (3Z)-cembrene A: diterpenes from a termite soldier (*Isoptera: Termitidae; Termitinae*).

33. See Mattern DL et al., Cembrene A and a congeneric ketone isolated from the paracloacal glands of the Chinese alligator (*Alligator sinensis*).

34. See Schulz S et al., New terpene hydrocarbons from the alligatoridae (crocodylia, reptilia).

35. See Rodríguez AD, The natural products chemistry of West Indian gorgonian octocorals.

36. See Ferchmin PA et al., Actions of octocoral and tobacco cembranoids on nicotinic receptors.

37. See Ferchmin PA et al., Tobacco cembranoids block behavioral sensitization to nicotine and inhibit neuronal acetylcholine receptor function.

38. Pagán OR, Effects of Cembranoids and Anesthetic Agents on [^{3}H]-Tenocyclidine Binding to the Nicotinic Acetylcholine Receptor from *Torpedo californica*.

39. Parthenolide was tested on the nicotinic receptor using two techniques, radioligand binding and electrophysiology. These latter experiments were done in collaboration with Dr. Gabriela Gonzalez-Bonet, a friend and Cornell classmate.

40. This actually reminds me of a quote attributed to one of the true polymaths of our time, Isaac Asimov: "The most exciting phrase to hear in science, the one that heralds new discoveries, is not 'Eureka!' (I found it!) but 'That's funny....'"

41. Incidentally, those two students are having great careers of their own. One of them is a neuroscience PhD, working on her postdoctoral training, and the other one is an MD in a neurology residence program. How proud am I of them?

42. Pagán OR et al., Toxicity and behavioral effects of dimethylsulfoxide in planaria. The very first "Pagán Lab" paper!

43. See Pagán OR et al., Reversal of cocaine-induced planarian behavior by parthenolide and related sesquiterpene lactones.

44. See Rowlands AL, Pagán OR, Parthenolide prevents the expression of cocaine-induced withdrawal behavior in planarians.

45. Schwarz D et al., Parthenolide blocks cocaine's effect on spontaneous firing activity of dopaminergic neurons in the ventral tegmental area. The paper that linked planarian and vertebrate neuropharmacology.

46. Pagán OR et al., A cembranoid from tobacco prevents the expression of nicotine-induced withdrawal behavior in planarian worms.

EPILOGUE

1. This story is entirely plausible. Charles Darwin did visit Hobart in 1836 during the voyage of the Beagle. As he did wherever he went, he collected specimens of the fauna and flora of the island. In his notes he does not mention collecting aquatic planarians, but he did record that he found specimens of a terrestrial planaria, which he named *Planaria tasmaniana* (later reclassified as *Tasmanoplana tasmaniana*). Furthermore, he conducted regeneration experiments on them. This organism has the distinction of being the only Tasmanian terrestrial animal discovered and named by Darwin himself. I'd like to think that he also saw freshwater planarians in Hobart.

Bibliography

BOOKS

Allen JS (2009) The Lives of the Brain: Human Evolution and the Organ of Mind. Cambridge: Belknap Press of Harvard University Press; 1st edition.
Bainbridge D (2010) Beyond the Zonules of Zinn: A Fantastic Journey Through Your Brain. Cambridge: Harvard University Press; 1st edition.
Boese A (2007) Elephants on Acid: And Other Bizarre Experiments. New York: Mariner Books; 1st edition. Chapter 3.
Bonner JT (2008) The Social Amoebae: The Biology of Cellular Slime Molds. Princeton: Princeton University Press.
Brøndsted HV (1969) Planarian Regeneration (International Series of Monographs in Pure and Applied Biology. Division: Zoology). Pergamon Press; 1st edition.
Brooks M (2001) Fly: The Unsung Hero of Twentieth Century Science. New York: Ecco Press.
Brown A (2003) In the Beginning Was the Worm. New York: Columbia University Press.
Byrne JH, Roberts JL (2003) From Molecules to Networks: An Introduction to Cellular and Molecular Neuroscience. San Diego: Academic Press; 1st edition.
Carolei A et al. (2008a) Catecholamines in Planaria. In: Raffa RB, Rawls SM, Editors. Planaria: A Model for Drug Action and Abuse. Austin, TX: Landes Bioscience, pp. 20–23.
Carolei A et al. (2008b) Opioids in Planaria. In: Raffa RB, Rawls SM, Editors. Planaria: A Model for Drug Action and Abuse. Austin, TX: Landes Bioscience, pp. 24–30.
Carroll PM, Fitzgerald K (2003) Model Organisms in Drug Discovery. Chichester: Wiley.
Chamovitz D (2012) What a Plant Knows: A Field Guide to the Senses. New York: Scientific American/Farrar, Straus and Giroux.
Colansanti A, Venturini G (2008) Nitric oxide in lower invertebrates. In: Raffa RB, Rawls SM, Editors. Planaria: A Model for Drug Action and Abuse. Austin, TX: Landes Bioscience, pp. 31–39.

Collins HM, Pinch T (2012) The Golem: What You Should Know About Science (Canto Classics). New York: Cambridge University Press; 2nd edition. Chapter 1.

Conniff R (2010) The Species Seekers: Heroes, Fools, and the Mad Pursuit of Life on Earth. New York: W. W. Norton & Company; 1st edition.

Corning WC, Ratner SC, Editors (1967) The Chemistry of Learning: Invertebrate Research. New York: Plenum Press.

Cowan MW et al. (2001) Synapses. Baltimore: The Johns Hopkins University Press.

Crick F (1990) What Mad Pursuit: A Personal View of Scientific Discovery. New York: Basic Books. Chapter 2.

Darwin CR (1882) The Various Contrivances by Which Orchids Are Fertilised by Insects. London: John Murray; 2nd edition. http://darwin-online.org.uk

Dalyell JG (1814) Observations on Some Interesting Phenomena in Animal Physiology, Exhibited by Several Species of Planariae: Illustrated by Coloured Figures of Living Animals. Printed by Andrew Balfour, for Archibald Constable & Co., Edinburgh, and Longman, Hurst, Rees, Orme and Brown, London.

Di Justo P, Grazier K (2010) The Science of Battlestar Galactica. Hoboken: Wiley; 1st edition.

Dunn R (2010) Every Living Thing: Man's Obsessive Quest to Catalog Life, from Nanobacteria to New Monkeys. New York: Harper Perennial; 1st edition.

Eisner T, Meinwald J, Editors (1995) Chemical Ecology: The Chemistry of Biotic Interaction. Washington, D.C.: National Academy Press.

Fields RD (2009) The Other Brain: From Dementia to Schizophrenia, How New Discoveries About the Brain Are Revolutionizing Medicine and Science. New York: Simon & Schuster; 1st edition.

Finger S (2004) Minds Behind the Brain: A History of the Pioneers and Their Discoveries. New York: Oxford University Press; 1st edition.

Firn R (2010) Nature's Chemicals: The Natural Products That Shaped Our World. Oxford: Oxford University Press.

Francis RC (2011) Epigenetics: The Ultimate Mystery of Inheritance. New York: W. W. Norton & Company.

Gould SJ (1994). Evolution in fact and theory. In: Hen's Teeth and Horse's Toes. New York: W. W. Norton & Company, pp. 253–62.

Hille B (2001) Ion Channels of Excitable Membranes. Sunderland: Sinauer; 3rd edition.

Holmes R (2010) The Age of Wonder: The Romantic Generation and the Discovery of the Beauty and Terror of Science. New York: Vintage; 1st edition.

Hyman LH (1940) The Invertebrates. Vol. 1, Protozoa Through Ctenophora. New York: McGraw-Hill.

Hyman LH (1951a) The Invertebrates. Vol. 2, Platyhelminthes and Rhynchocoela. New York: McGraw-Hill.

Hyman LH (1951b) The Invertebrates. Vol. 3, Acanthocephala, Aschelminthes, and Entoprocta. New York: McGraw-Hill.

Hyman LH (1955) The Invertebrates. Vol. 4, Echinodermata. New York: McGraw-Hill.

Hyman LH (1959) The Invertebrates. Vol. 5, Smaller Coelomate Groups. New York: McGraw-Hill.

Hyman LH (1967) The Invertebrates. Vol. 6, Molusca I. New York: McGraw-Hill.

Iversen L et al. (2008) Introduction to Neuropsychopharmacology. New York: Oxford University Press; 1st edition.

Jenkins MM (1967) Aspects of planarian biology and behavior. In: Corning WC, Ratner SC, Editors. The Chemistry of Learning: Invertebrate Research. New York: Plenum Press, pp. 116–43.

Jones S (2011) The Darwin Archipelago: The Naturalist's Career Beyond Origin of Species. Ann Arbor: Yale University Press; 1st edition.
Kauffman S (1995) At Home in the Universe: The Search for the Laws of Self-Organization and Complexity. New York: Oxford University Press.
Kaplan R (2001) Science Says. New York: Freeman, p. 203
Koch C (2012) Consciousness-Confessions of a Romantic Reductionist. Cambridge: MIT Press.
Larsen L (2005) In All Their Animal Brilliance: Poems. Tampa: University of Tampa Press.
Lentz TL (1968) Primitive Nervous Systems. New Haven: Yale University Press; 1st edition.
Llinás RR (1999) The Squid Giant Synapse. A Model for Chemical Transmission. New York: Oxford University Press.
Maczulak A (2010) Allies and Enemies: How the World Depends on Bacteria. Upper Saddle River: FT Press; 1st edition.
Margulis L (1999) Symbiotic Planet: A New Look at Evolution. New York: Basic Books.
Margulis L, Sagan D (2003) Acquiring Genomes: A Theory of the Origins of the Species. New York: Basic Books.
Margulis L, Sagan D (2000) What Is Life? Berkeley: University of California Press.
McConnell JV, Schutjer M (1971) Science, Sex, and Sacred Cows: Spoofs on Science from the Worm Runner's Digest. New York. Harcourt Brace Jovanovich; 1st edition,
Marshall LH, Magoun HW (1998) Discoveries in the Human Brain: Neuroscience Prehistory, Brain Structure, and Function. Totowa: Humana Press; 1st edition.
Mather JA et al. (2010) Octopus: The Ocean's Intelligent Invertebrate. Portland: Timber Press; 1st edition.
Mayr E (2002) What Evolution Is. New York: Basic Books.
Mazzarello P (2009) Golgi: A Biography of the Founder of Modern Neuroscience. New York: Oxford University Press; 1st edition.
Miles FA (1969) Excitable Cells (Monographs in Modern Biology for Upper School and University Courses). Charlotte: Heinemann Medical.
Mitman G, Fausto-Sterling A (1992) Whatever happened to planaria? C.M. Child and the physiology of inheritance. In: Clarke AE, Fujimura, Editors. The Right Tools for the Job. Princeton: Princeton University Press, pp. 172–97.
Newman L, Cannon L (2003) Marine Flatworms: The World of Polyclads. Collingwood: CSIRO Publishing.
Nishimura K et al. (2008) Brain and neural networks. In: Raffa RB, Rawls SM, Editors. Planaria: A Model for Drug Action and Abuse. Austin, TX: Landes Bioscience, pp.4–12.
Nusslein-Volhard C, Dahm R (2002) Zebrafish: A Practical Approach. New York: Oxford University Press.
O'Shea M (2006) The Brain: A Very Short Introduction. New York: Oxford University Press; 1st edition.
Pechenik JA (2005) Biology of the Invertebrates. New York: McGraw-Hill; 5th edition.
Punset E (2011) Excusas para no pensar. Barcelona: Editorial Planeta, pp. 9–13.
Raffa RB et al. (2004) Netter's Illustrated Pharmacology (Netter Basic Science). Saunders; 1st edition.
Raffa RB, Rawls SM (2008) Planaria: A Model for Drug Action and Abuse. Austin, TX: Landes Bioscience.
Ramón y Cajal S (1917) Recuerdos de mi vida. Tomo II—Historia de mi Labor Científica. Madrid, España: Imprenta y Librería de Nicolás Moya.
Rapport R (2005) Nerve Endings: The Discovery of the Synapse. New York: Norton & Company.

Ruppert EE, Fox RS, Barnes RD (2003) Invertebrate Zoology: A Functional Evolutionary Approach. Independence: Brooks Cole; 7th edition, pp. 197–233.

Seung S (2012) Connectome. New York: Houghton Mifflin Harcourt Trade.

Snyder LJ (2012) The Philosophical Breakfast Club: Four Remarkable Friends Who Transformed Science and Changed the World. New York: Broadway Books.

Stewart I (2011) The Mathematics of Life. New York: Basic Books; 1st edition.

Stockwell BR (2011) The Quest for the Cure: The Science and Stories Behind the Next Generation of Medicines. New York: Columbia University Press. Chapter 1.

Stott R (2012) Darwin's Ghosts: The Secret History of Evolution. New York: Spiegel & Grau.

Streatfeild D (2002) Cocaine: An Unauthorized Biography. New York: Thomas Dunne Books; 1st edition.

Strogatz SH (2004) Sync: How Order Emerges From Chaos in the Universe, Nature, and Daily Life. New York: Hyperion; Reprint edition.

Swanson LW (2011) Brain Architecture: Understanding the Basic Plan. New York: Oxford University Press; 2nd edition.

Tanford C, Reynolds J (2004) Nature's Robots: A History of Proteins. Oxford: Oxford University Press.

Tompkins P, Bird C (1989) The Secret Life of Plants: A Fascinating Account of the Physical, Emotional, and Spiritual Relations Between Plants and Man. New York: Harper & Row.

Venturini G et al. (1986) Peptide-monoamine interactions in planaria and hydra. In: Stefano GB, Editor. Handbook of Comparative Opioid and Related Neuropeptide Mechanisms. Boca Raton, FL: CRC Press, pp. 245–54.

Weiner J (1999) Time, Love, Memory: A Great Biologist and His Quest for the Origins of Behavior. New York: Knopf.

Wells PH (1967) Training planarians in a van Oye maze. In: Corning WC, Ratner SC, Editors. The Chemistry of Learning: Invertebrate Research. New York: Plenum Press, pp. 251–54.

Wharton DA (2002) Life at the Limits: Organisms in Extreme Environments. Cambridge: Cambridge University Press.

Williams W (2011) Kraken: The Curious, Exciting, and Slightly Disturbing Science of Squid. New York: Abrams Image.

Wilson EO (2003) The Future of Life. New York: Vintage Books, p. 120.

Yoon CK (2009) Naming Nature: The Clash Between Instinct and Science. New York: W. W. Norton & Company; 1st edition.

Zimmer C (2005) Soul Made Flesh: The Discovery of the Brain—And How It Changed the World. New York: Free Press.

ARTICLES

Aboobaker AA (2011) Planarian stem cells: a simple paradigm for regeneration. Trends Cell Biol. 21(5):304–11.

Aboobaker AA, Kao D (2012) A lack of commitment for over 500 million years: conserved animal stem cell pluripotency. EMBO J. 31(12):2747–49.

Achatz JG et al. (2012) The Acoela: on their kind and kinships, especially with nemertodermatids and xenoturbellids (Bilateria incertae sedis). Org Divers Evol. doi:10.1007/s13127-012-0112-4

Achatz JG, Martinez P (2012) The nervous system of Isodiametra pulchra (Acoela) with a discussion on the neuroanatomy of the Xenacoelomorpha and its evolutionary implications. Front Zool. 9(1):27. doi:10.1186/1742-9994-9-27

Adams ED et al. (2010) Freshwater sponges have functional, sealing epithelia with high transepithelial resistance and negative transepithelial potential. PLoS One. 5(11):e15040.

Agata K et al. (1998) Structure of the planarian central nervous system (CNS) revealed by neuronal cell markers. Zoolog Sci. 15(3):433–40.

Agata K et al. (2006) Two different evolutionary origins of stem cell systems and their molecular basis. Semin Cell Dev Biol. 17(4):503–9.

Agata K et al., (2007) Unifying principles of regeneration I: Epimorphosis versus morphallaxis. Dev Growth Differ 49(2):73–8.

Agata K, Inoue T (2012) Survey of the differences between regenerative and non-regenerative animals. Dev Growth Differ. 54(2):143–52.

Agata K, Watanabe K (1999) Molecular and cellular aspects of planarian regeneration. Semin Cell Dev Biol. 10(4):377–83.

Agnati LF et al. (2007) One century of progress in neuroscience founded on Golgi and Cajal's outstanding experimental and theoretical contributions. Brain Res Rev. 55(1):167–89.

Ahn AC et al. (2006) The limits of reductionism in medicine: could systems biology offer an alternative? PLoS Med. 3(6):e208.

Algeri S et al. (1983) Effects of dopaminergic agents on monoamine levels and motor behaviour in planaria. Comp Biochem Physiol C. 74(1):27–29.

Alivisatos AP et al. (2012) The brain activity map project and the challenges of functional connectomics. Neuron 74:970–74.

Allison CW (1975) Primitive fossil flatworm from Alaska: new evidence bearing on ancestry of the Metazoa. Geology. 3(11):649–52.

Almuedo-Castillo M et al. (2012) Wnt signaling in planarians: new answers to old questions. Int J Dev Biol. 56(1–3):53–65.

Alpi A et al. (2007) Plant neurobiology: no brain, no gain? Trends Plant Sci. 12(4):135–36.

Alves RR, Alves HN (2011) The faunal drugstore: animal-based remedies used in traditional medicines in Latin America. J Ethnobiol Ethnomed. 7:9.

Anonymous (1891) Microscopy. The Am Nat. 25(296):771–72.

Aoki R et al. (2009) Recording and spectrum analysis of the planarian electroencephalogram. Neuroscience. 159(2):908–14.

Arendt D et al. (2008) The evolution of nervous system centralization. Philos Trans R Soc Lond B Biol Sci. 363(1496):1523–28.

Azevedo FA et al. (2009) Equal numbers of neuronal and nonneuronal cells make the human brain an isometrically scaled-up primate brain. J Comp Neurol. 513(5):532–41.

Baguñà J (2012) The planarian neoblast: the rambling history of its origin and some current black boxes. Int J Dev Biol. 56(1–3):19–37.

Baguñà J et al. (2008) Back in time: a new systematic proposal for the Bilateria. Philos Trans R Soc Lond B Biol Sci. 363(1496):1481–91.

Bailly X et al. (2013) The urbilaterian brain revisited: novel insights into old questions from new flatworm clades. Dev Genes Evol. 223(3):149–57.

Baker D et al. (2011) Minimal structural requirements of alkyl Y-lactones capable of antagonizing the cocaine-induced motility decrease in planarians. Pharmacol Biochem Behav. 100(1):174–79.

Baldwin MR, Barbieri JT (2009) Association of botulinum neurotoxins with synaptic vesicle protein complexes. Toxicon. 54(5):570–74.

Baluška F et al. (2004) Root apices as plant command centres: the unique 'brain-like' status of the root apex transition zone. Biologia (Bratisl). 59:9–17.

Baluska F, Mancuso S (2007) Plant neurobiology as a paradigm shift not only in the plant sciences. Plant Signal Behav. 2(4):205–7.

Bargmann CI (2012) Beyond the connectome: how neuromodulators shape neural circuits. Bioessays. 34(6):458–65.

Bargmann CI, Marder E (2013) From the connectome to brain function. Nat Methods. 10(6):483–90.

Bauman J, DiDomenico R (2002) Cocaine-induced channelopathies: emerging evidence on the multiple mechanisms of sudden death. Cardiovasc Pharmacol Ther. 7(3):195–202.

Bely AE (2010) Evolutionary loss of animal regeneration: pattern and process. Integr Comp Biol. 50(4):515–27.

Bely AE, Nyberg KG (2010) Evolution of animal regeneration: re-emergence of a field. Trends Ecol Evol. 25(3):161–70.

Bely AE, Sikes JM (2010) Acoel and platyhelminth models for stem-cell research. J Biol. 9(2):14.

Benazzi M et al. (1975) Further contribution to the taxonomy of the *Dugesia lugubris-polychroa* group with description of *Dugesia mediterranea* n. sp. (Tricladida, Paludicola). Bolletino di zoologia 42(1):81–89.

Bennett EL, Calvin M (1964) Failure to train planarians reliably. Neurosci Res Program Bull. July–August:3–14.

Bennett MR (1999) The early history of the synapse: from Plato to Sherrington. Brain Res Bull. 50(2):95–118.

Best JB (1963) Protopsychology. Sci Am. 208:54–62.

Best JB, Morita M (1991) Toxicology of planarians. Hydrobiologia. 227(1):375–83.

Birnbaum KD, Sánchez Alvarado A (2008) Slicing across kingdoms: regeneration in plants and animals. Cell. 132(4):697–710.

Blackiston D et al. (2010) A second-generation device for automated training and quantitative behavior analyses of molecularly-tractable model organisms. PLoS One. 5(12):e14370. doi:10.1371/journal.pone.0014370

Boeckxstaens GE (2002) Understanding and controlling the enteric nervous system. Best Pract Res Clin Gastroenterol. 16(6):1013–23.

Bonfanti L (2011) From hydra regeneration to human brain structural plasticity: a long trip through narrowing roads. ScientificWorldJournal. 11:1270–99.

Bourlat SJ, Hejnol A (2009) Acoels. Curr Biol. 19(7):R279–80.

Bower JM, Parsons LM (2003) Rethinking the "lesser brain." Sci Am. 289(2):50–57.

Brenner ED et al. (2006) Plant neurobiology: an integrated view of plant signaling. Trends Plant Sci. 8:413–19.

Brenner S (2009) In the beginning was the worm. Genetics. 182(2):413–15.

Brodie ED 3rd (2009) Toxins and venoms. Curr Biol. 19(20):R931–35.

Bullock TH et al. (2005) Neuroscience. The neuron doctrine, redux. Science. 310(5749):791–93.

Buttarelli FR et al. (2000) Acetylcholine/dopamine interaction in planaria. Comp Biochem Physiol C Toxicol Pharmacol. 125(2):225–31.

Buttarelli FR et al. (2002) Cannabinoid-induced stimulation of motor activity in planaria through an opioid receptor-mediated mechanism. Prog Neuropsychopharmacol Biol Psychiatry. 26(1):65–68.

Buttarelli FR et al. (2008) Neuropharmacology and behavior in planarians: translations to mammals. Comp Biochem Physiol C Toxicol Pharmacol. 147(4):399–408.

Calahorro F, Ruiz-Rubio M (2011) Caenorhabditis elegans as an experimental tool for the study of complex neurological diseases: Parkinson's disease, Alzheimer's disease and autism spectrum disorder. Invert Neurosci. 11(2):73–83.

Calatayud J, González A (2003) History of the development and evolution of local anesthesia since the coca leaf. Anesthesiology. 98(6):1503–8.

Callier S et al. (2003) Evolution and cell biology of dopamine receptors in vertebrates. Biol Cell. 95(7):489–502.

Carew TJ, Sahley CL (1986) Invertebrate learning and memory: from behavior to molecules. Ann Rev Neurosci. 9:23–28.

Carolei A et al. (1975) Proposal of a new model with dopaminergic-cholinergic interactions for neuropharmacological investigations. Neuropsychobiology. 1(6):355–64.

Caronti B et al. (1999) Treatment with 6-hydroxydopamine in planaria (Dugesia gonocephala s.l.): morphological and behavioral study. Comp Biochem Physiol C Pharmacol Toxicol Endocrinol. 123(3):201–7.

Cebrià F (2007) Regenerating the central nervous system: how easy for planarians! Dev Genes Evol. 217(11–12):733–48.

Cebrià F et al. (1996) Myosin heavy chain gene in Dugesia (G.) tigrina: a tool for studying muscle regeneration in planarians. Int J Dev Biol. Suppl 1:177S–78S.

Cebrià F et al. (2002a) FGFR-related gene nou-darake restricts brain tissues to the head region of planarians. Nature. 419(6907):620–24.

Cebrià F et al. (2002b) The expression of neural-specific genes reveals the structural and molecular complexity of the planarian central nervous system. Mech Dev. 116(1–2):199–204.

Cebrià F et al. (2002c) Dissecting planarian central nervous system regeneration by the expression of neural-specific genes. Dev Growth Differ. 44(2):135–46.

Ciereszco LS (1989) Sterol and diterpenoid production by Zooxanthellae in coral reefs: a review. Biol Oceanogr. 6:363–74.

Cieresziko LS et al. (1960) Chemistry of coelenterates. I. Occurrence of terpenoid compounds in gorgonians. Ann N Y Acad Sci. 90:917–19.

Cloud P et al. (1976) Traces of animal life from 620-million-year-old rocks in North Carolina. Am Sci. 64(4):396–406.

Cohen JE (2004) Mathematics is biology's next microscope, only better; biology is mathematics' next physics, only better. PLoS Biol. 2(12):e439.

Cook HC (1997) Origins of tinctorial methods in histology. J Clin Pathol. 50(9):716–20.

Corning WC, John ER (1961) Effect of ribonuclease on retention of conditioned response in regenerated planarians. Science. 134(3487):1363–65.

Culver P, Jacobs RS (1981) Lophotoxin: a neuromuscular acting toxin from the sea whip (Lophogorgia rigida). Toxicon. 19(6):825–30.

Davies L et al. (1985) Nerve repair and behavioral recovery following brain transplantation in Notoplana acticola, a polyclad flatworm. J Exp Zool. 235(2):157–73.

Davies PC (2011) Searching for a shadow biosphere on Earth as a test of the 'cosmic imperative'. Philos Transact A Math Phys Eng Sci. 369(1936):624–32.

Davies PC et al. (2009) Signatures of a shadow biosphere. Astrobiology. 9(2):241–49.

Dawkins R, Krebs JR (1979) Arm races between and within species. Proc R Soc Lond B. 205:489–511.

DeCarlos JA, Borrell J (2007) A historical reflection of the contributions of Cajal and Golgi to the foundations of neuroscience. Brain Res Rev. 55:8–16.

DeFelipe J (2002) Sesquicentenary of the birthday of Santiago Ramón y Cajal, the father of modern neuroscience. Trends Neurosci. 25(9):481–84.

DeFelipe J (2006) Brain plasticity and mental processes: Cajal again. Nat Rev Neurosci. 7:811–18.

DeFelipe J (2010) From the connectome to the synaptome: an epic love story. Science. 330(6008):1198–201.

Denes AS et al. (2007) Molecular architecture of annelid nerve cord supports common origin of nervous system centralization in bilateria. Cell. 129(2):277–88.

Dentzien-Dias PC et al. (2012) Tapeworm eggs in a 270 million-year-old shark coprolite. PLoS One. 8(1):e55007.

Donato M (2010) Why is the language of nature mathematical? In Antonini G, Altamore A, Editors. Galileo and the Renaissance Scientific Discourse. Edizioni Nuova Cultura, Roma, pp. 65–71.

Eckert R, Brehm P (1979) Ionic mechanisms of excitation in Paramecium. Annu Rev Biophys Bioeng. 8:353–83.

Egger B (2006) The regeneration capacity of the flatworm *Macrostomum lignano*—on repeated regeneration, rejuvenation, and the minimal size needed for regeneration. Dev Genes Evol. 216(10):565–77.

Egger B (2008) Regeneration: rewarding, but potentially risky. Birth Defects Res C Embryo Today. 84(4):257–64.

Egger B et al. (2007) Free-living flatworms under the knife: past and present. Dev Genes Evol. 217(2):89–104.

Egger B et al. (2009) To be or not to be a flatworm: the acoel controversy. PLoS One. 4(5):e5502.

Ehrenberg CG (1831) Animalia evertebrata exclusis insectis recensuit Dr C.G. Ehrenberg. In Ehrenberg, Hemprich, Editors. Symbolae Physicae Berlin.

Elliott GR, Leys SP (2010) Evidence for glutamate, GABA and NO in coordinating behaviour in the sponge, Ephydatia muelleri (Demospongiae, Spongillidae). J Exp Biol. 213(Pt 13):2310–21.

Elliot SA, Sánchez Alvarado A (2012) The history and enduring contributions of planarians to the study of animal regeneration. WIREs Developmental Biology. doi:10.1002/wdev.82

Elrich PR, Raven PH (1967) Butterflies and plants. Sci Am. June. 216:104–13

Endo A (2010) A historical perspective on the discovery of statins. Proc Jpn Acad Ser B Phys Biol Sci. 86(5):484–93.

Erwin DH, Davidson EH (2002) The last common bilaterian ancestor. Development. 129:3021–32.

Essack M et al. (2012) Conotoxins that confer therapeutic possibilities. Mar Drugs. 10(6):1244–65.

Fenical W et al. (1981) Lophotoxin: a novel neuromuscular toxin from Pacific sea whips of the genus *Lophogorgia*. Science. 212(4502):1512–14.

Ferchmin PA et al. (2001) Tobacco cembranoids block behavioral sensitization to nicotine and inhibit neuronal acetylcholine receptor function. J Neurosci Res. 64:18–25.

Ferchmin PA et al. (2009) Actions of octocoral and tobacco cembranoids on nicotinic receptors. Toxicon. 54(8):1174–82.

Finnery JR (2003) The origins of axial patterning in the metazoa: how old is bilateral symmetry? Int J Dev Biol. 47:523–29.

Forterre P, Philippe H (1999) The last universal common ancestor (LUCA), simple or complex? Biol Bull. 196:373–77.

Fries EF (1928) Drug action in galvanotropic responses. J Gen Physiol. 11(5):507–13.

Fuqua WC et al. (1994) Quorum sensing in bacteria: the LuxR-LuxI family of cell density-responsive transcriptional regulators. J Bacteriol. 176(2):269–75.

Fuqua WC et al. (1996) Census and consensus in bacterial ecosystems: the LuxR-LuxI family of quorum-sensing transcriptional regulators. Annu Rev Microbiol. 50:727–51.

Gainetdinov R, Caron M (2003) Monoamine transporters: from genes to behavior. Annu Rev Pharmacol Toxicol 43:261–84.

Galliot B (2006) Regeneration in hydra. eLS. John Wiley & Sons Ltd.http://www.els.net. doi:10.1038/npg.els.0004186

Galliot B (2012) Hydra, a fruitful model system for 270 years. Int J Dev Biol. 56(6–8):411–23.

Galliot B, Quiquand M (2011) A two-step process in the emergence of neurogenesis. Eur J Neurosci. 34(6):847–62.

Garzón FC (2007) The quest for cognition in plant neurobiology. Plant Signal Behav. 2(4):208–11.

Gatherer D (2010) So what do we really mean when we say that systems biology is holistic? BMC Syst Biol. 4:22.

Gentile L et al. (2011) The planarian flatworm: an in vivo model for stem cell biology and nervous system regeneration. Dis Model Mech. 4(1):12–19.

Glauber KM et al. (2010) Hydra. Curr Biol. 20(22):R964–65.

Glickstein M (2006) Golgi and Cajal: the neuron doctrine and the 100th anniversary of the 1906 Nobel Prize. Curr Biol. 16(5):R147–51.

Golgi C (1906) Nobel lecture. http://www.nobelprize.org/nobel_prizes/medicine/laureates/1906/golgi-lecture.pdf

Goodenough DA, Paul DL (2009) Gap junctions. Cold Spring Harb Perspect Biol. 1(1):a002576. http://cshperspectives.cshlp.org/

Gross CG (1995) Aristotle on the brain. Neuroscientist. 1(4):245–50.

Gross CG (1998) Galen and the squealing pig. Neuroscientist. 4(3):216–21.

Guillery RW (2005) Observations of synaptic structures: origins of the neuron doctrine and its current status. Philos Trans R Soc Lond B Biol Sci. 360(1458):1281–307.

Guillery RW (2007) Relating the neuron doctrine to the cell theory. Should contemporary knowledge change our view of the neuron doctrine? Brain Res Rev. 55(2):411–21.

Gurley KA et al. (2008) Beta-catenin defines head versus tail identity during planarian regeneration and homeostasis. Science. 319(5861):323–27.

Gustafsson MKS, Reuter M (1990) The early brain: Proceedings of the symposium "Invertebrate Neurobiology" Abo Akademi, Finland, 1–178.

Hagen EH et al. (2009) Ecology and neurobiology of toxin avoidance and the paradox of drug reward. Neuroscience. 160(1):69–84.

Hagmann P (2005) From Diffusion MRI to Brain Connectomics. PhD dissertation. Ecole Polytechnique Fédérale de Lausanne, Lausanne.

Halai R, Craik DJ (2009) Conotoxins: natural product drug leads. Nat Prod Rep. 26(4):526–36.

Hann RM et al. (1998) Characterization of cembranoid interaction with the nicotinic acetylcholine receptor. J Pharmacol Exper Ther. 287:253–60.

Hann MM, Keserü GM (2012) Finding the sweet spot: the role of nature and nurture in medicinal chemistry. Nat Rev Drug Discov. 11(5):355–65.

Heath H (1907) A new turbellarian from Hawaii. Proc Acad Nat Sci Philadelphia. 59(1):145–48.

Hertting G, Axelrod J (1961) Fate of tritiated noradrenaline at the sympathetic nerve endings. Nature. 192:172–73.

Holstein TW et al. (2003) Cnidarians: an evolutionarily conserved model system for regeneration? Dev Dyn. 226(2):257–67.

Huffmann MA (1997) Current evidence for self-medication in primates: a multidisciplinary perspective. Yearbook Phys Anthropol. 40:171–200.

Hyman LH, Evelyn Hutchinson G (1991) Libbie Hyman 1888-1969. A biographical memoir. National Academy of Sciences. http://www.nasonline.org/publications/biographical-memoirs/memoir-pdfs/hyman-libbie.pdf

Ichikawa A, Kawakatsu M (1964) A new freshwater planarian *Dugesia japonica*, commonly but erroneously known as *Dugesia gonocephala* (Dugès). Annot Zool Japon. 37:185–94.

Jones EG (1999) Golgi, Cajal and the neuron doctrine. J Hist Neurosci. 8(2):170–78.

Jones EG (2011) Cajal's debt to Golgi. Brain Res Rev. 66(1–2):83–91.

Kaun KR et al. (2011) A Drosophila model for alcohol reward. Nat Neurosci. 14(5):612–19.

Kawakatsu M (1942) [Collection and observation of plants in the vicinity of Sonobe-chô, Kyôto Prefecture]. Enjyô. 14:10–11 (in Japanese).

Kawakatsu M (1965) On the ecology and distribution of freshwater planarians in the Japanese Islands, with special references to their vertical distribution. Hydrobiologia. 26:349–408.

Kawakatsu M (1968) North American triclad Turbellaria, 17: freshwater planarians from Lake Tahoe. Proc U S Nat Mus. 124(3638):1–21.

Kawakatsu M et al. (1982) *Bipalium nobile* sp.nov. (Turbellaria,Tricladida,Terricola), a New Land Planarian from Tokyo. Annot Zool Japon. 55(4):236–62.

Keesey J (2005) How electric fish became sources of acetylcholine receptor. J Hist Neurosci. 14(2):149–64.

Kenk R (1941) A fresh-water triclad from Puerto Rico, *Dugesia antillana*, new species. Occasional papers of the Museum of Zoology University of Michigan. 436:1–9.

Kerney R et al. (2011) Intracellular invasion of green algae in a salamander host. Proc Natl Acad Sci U S A. 108(16):6497–502.

Kingston DG (2011) Modern natural products drug discovery and its relevance to biodiversity conservation. J Nat Prod. 74(3):496–511.

Komai T, Kawakatsu M (1970) In memoriam - Dr. Libbie H. Hyman. J Biol Psychol/Worm Runner's Digest 12(1):3–23

Koopowitz H, Holman M (1988) Neuronal repair and recovery of function in the polyclad flatworm, *Notoplana acticola*. Am Zool. 28:1065–75.

Kritsky G (1991). Darwin's Madagascan Hawk Moth prediction. Am Entomol. 37:206–9.

Ladurner P et al. (2005) A new model organism among the lower Bilateria and the use of digital microscopy in taxonomy of meiobenthic Platyhelminthes: *Macrostomum lignano*, n. sp. (Rhabditophora, Macrostomorpha). JZS. 43(2):114–26.

Lederman LM (1984) The value of fundamental science. Sci Am. 251(5):40–47.

Lenicque PM et al. (1972) Effects of some components of *Cannabis sativa* on the regenerating planarian worm *Dugesia tigrina*. Experientia. 28(11):1399–400.

Li YH, Tian X (2012) Quorum sensing and bacterial social interactions in biofilms. Sensors (Basel). 12(3):2519–38.

Liu SY et al. (2013) Reactivating head regrowth in a regeneration-deficient planarian species. Nature 500(7460):81–4.

Lobo D et al. (2012) Modeling planarian regeneration: a primer for reverse-engineering the worm. PLoS Comput Biol. 8(4):e1002481.

Margotta V et al. (1975) Osservazioni morfologiche ed istochimiche sui recettori sinaptici di Dugesia gonocephala s.l., in varie condizioni sperimentali. Rend Acc Naz Lincei Serie VIII. 59:201–9.

Margotta V et al. (1976) Il ruolo del cervello di Dugesia gonocephala s.l. nella risposta motoria a stimoli. Rend Acc Naz Lincei Serie VIII. 61:133–42.

Margotta V et al. (1979) Action of several food colours on whole and regenerant Planaria. Riv Biol. 72(1–2):11–38.

Margotta V et al. (1997) Effects of cocaine treatment on the nervous system of planaria (Dugesia gonocephala s. l.). Histochemical and ultrastructural observations. Eur J Histochem. 41(3):223–30.

Markram H (2012) The human brain project. Sci Am. 306(6):50–55.

Martín-Durán JM et al. (2012) Planarian embryology in the era of comparative developmental biology. Int J Dev Biol. 56(1–3):39–48.

Martínez DE (1998) Mortality patterns suggest lack of senescence in hydra. Exp Gerontol. 33:217–25.

Matheson T (2002) Invertebrate nervous systems. eLS. John Wiley & Sons. http://www.els.net. doi:10.1038/npg.els.0003637

Mattern DL et al. (1997) Cembrene A and a congeneric ketone isolated from the paracloacal glands of the Chinese alligator (*Alligator sinensis*). J Nat Prod. 60(8):828–31.

Mayer EA (2011) Gut feelings: the emerging biology of gut–brain communication. Nat Rev Neurosci. 12(8):453–66.

Maxmen A (2011) A can of worms. Nature. 470:161–62.

McConnell JV (1966) Comparative physiology: learning in invertebrates. Annu Rev Physiol. 28:107–36.

McConnell JV (1964) Cannibalism and memory in flatworms. New Scientist. 21:465–68.

McConnell JV, Editor (1965) A manual of psychological experimentation on planarians. Special publication of The Worm Runner's Digest.

McConnell JV et al. (1958) The effects of regeneration upon retention of a conditioned response in the planarian. J Comp Physiol Psychol. 52(1):1–5.

Micheva KD et al. (2010) Single-synapse analysis of a diverse synapse population: proteomic imaging methods and markers. Neuron. 68(4):639–53.

Miller JA (1938) Studies on heteroplastic transplantation in triclads I. Cephalic grafts between *Euplanaria dorotocephala* and *E. tigrina*. Physiol Zool. 11(2):214–47.

Mineta K et al. (2003) Origin and evolutionary process of the CNS elucidated by comparative genomics analysis of planarian ESTs. Proc Natl Acad Sci U S A. 100(13):7666–71.

Montgomery JR, Coward SJ (1974) On the minimal size of a planarian capable of regeneration. Trans Amer Micros Soc. 93:386–91.

Moore R (2001) The "rediscovery" of Mendel's work. Bioscene. 27(2):13–24.

Moraczewski J et al. (2008) From planarians to mammals - the many faces of regeneration. Int J Dev Biol 52(2–3):219–27.

Morgan TH (1898) Experimental studies of the regeneration of *Planaria maculata*. Arch Entw Mech Org. 7:364–97.

Moroz LL (2009) On the independent origins of complex brains and neurons. Brain Behav Evol. 74(3):177–90.

Morth JP et al. (2011) A structural overview of the plasma membrane Na+,K+-ATPase and H+-ATPase ion pumps. Nat Rev Mol Cell Biol. 12(1):60–70.

Murakoshi H, Yasuda R (2011) Postsynaptic signaling during plasticity of dendritic spines. Trends Neurosci. 35(2):135–43.

Nakagaki T (2001) Smart behavior of true slime mold in a labyrinth. Res Microbiol. 152(9):767–70.

Nakagaki T et al. (2000) Maze-solving by an amoeboid organism. Nature. 407(6803):470.

Nasser M (1987) Psychiatry in ancient Egypt. Psychiatric Bull. 11:420–22.

Nebel A, Bosch TC (2012) Evolution of human longevity: lessons from Hydra. Aging. 4(11):730–1.

Needleman HL (1967) Tolerance and dependence in the planarian after continuous exposure to morphine. Nature. 215(5102):784–85.

Needleman HL (2008) The case of Deborah Rice: who is the Environmental Protection Agency protecting? PLoS Biol. 6(5):e129.

Netsky MG (1986) What is a brain, and who said so? Br Med J (Clin Res Ed). 293(6562):1670–72.

Niehrs C (2010) On growth and form: a Cartesian coordinate system of Wnt and BMP signaling specifies bilaterian body axes. Development. 137(6):845–57.

Nielsen R et al. (2005) A scan for positively selected genes in the genomes of humans and chimpanzees. PLoS Biol. 3(6):170.

Nimchinsky EA et al. (2002) Structure and function of dendritic spines. Annu Rev Physiol. 64:313–53.

Northcutt RG (2012) Evolution of centralized nervous systems: two schools of evolutionary thought. Proc Natl Acad Sci U S A. 109(Suppl 1):10626–33.

Okugawa, KI (1957) An experimental study of sexual induction in the asexual form of Japanese fresh-water planarian, *Dugesia gonocephala* (Dugès). Bull Kyoto Gakugei Univ B. 11:8–27.

Olivera BM, Cruz L (2001) Conotoxins, in retrospect. Toxicon. 39:7–14.

Oviedo NJ, Beane WS (2009) Regeneration: The origin of cancer or a possible cure? Semin Cell Dev Biol. 20(5):557–64.

Pagán OR (1998) Effects of Cembranoids and Anesthetic Agents on [3H]-Tenocyclidine Binding to the Nicotinic Acetylcholine Receptor from *Torpedo californica*. MS thesis. University of Puerto Rico, San Juan, Puerto Rico.

Pagán OR et al. (2001) Cembranoid and long-chain alkanol sites on the nicotinic acetylcholine receptor and their allosteric interaction. Biochemistry. 40(37):11121–30.

Pagán OR (2005) Synthetic Local Anesthetics as Alleviators of Cocaine Inhibition of the Human Dopamine Transporter. PhD dissertation. Cornell University, Ithaca, NY.

Pagán OR et al. (2006) Toxicity and behavioral effects of dimethylsulfoxide in planaria. Neurosci Lett. 407(3):274–78.

Pagán OR et al. (2008) Reversal of cocaine-induced planarian behavior by parthenolide and related sesquiterpene lactones. Pharmacol Biochem Behav. 89(2):160–70.

Pagán OR et al. (2009a) A cembranoid from tobacco prevents the expression of nicotine-induced withdrawal behavior in planarian worms. Eur J Pharmacol. 615(1–3):118–24.

Pagán OR et al. (2009b) The flatworm planaria as a toxicology and behavioral pharmacology animal model in undergraduate research experiences. J Undergrad Neurosci Educ. 7(2):A48–52. http://www.funjournal.org/

Pagán OR et al. (2012) Planarians in pharmacology: parthenolide is a specific behavioral antagonist of cocaine in the planarian *Girardia tigrina*. Int J Dev Biol. 56(1–3):193–6.

Pagán OR et al. (2013) Planarians require an intact brain to behaviorally react to cocaine, but not to react to nicotine. Neuroscience. 246:265–70.

Palladini G et al. (1976a) Effect of some environmental pollutants on the behaviour of motility of *Dugesia gonocephala* S.L. Note I. Riv Biol. 69(3–4):155–67.

Palladini G et al. (1976b) Pharmacological, electronmicroscopic and histochemical study on an in-vivo model of the dopaminergic nervous system (*Dugesia gonocephala* s.l). Acta Neurol (Napoli). 31(1):1–5 (in Italian).

Palladini G et al. (1976c) Studio sulla sensibilità al rame dei neuroni dopaminergici: azione sul sistema nervoso di *Dugesia gonocephala* s.l. Rend Acc Naz Lincei Serie VIII. 60:523–28.

Palladini G et al. (1977a) Azione dopaminergica del colorante alimentare E-102 (tartrazina), nella planaria e nel topo. Rivista di Tossicologia Sperimentale e Clinica: 1–2:93–100.

Palladini G et al. (1977b) Studio sulla sensibilità allo zinco del sistema nervoso di *Dugesia gonocephala* s.l. Confronto con l'azione del rame e di altri metalli pesanti. Rend Acc Naz Lincei Serie VIII. 63:447–53.

Palladini G et al. (1978) Effetti di soluzioni di diversi anioni a varia pressione osmotica e viscosità sul sistema nervoso dopaminergico di *Dugesia gonocephala* s.l. Riv Biol. 71(1–4):41–62.

Palladini G et al. (1979a) The pigmentary system of planaria. I. Morphology. Cell Tissue Res. 199(2):197–202.

Palladini G et al. (1979b) The pigmentary system of planaria. II. Physiology and functional morphology. Cell Tissue Res. 199(2):203–11.

Palladini G et al. (1980) Dopamine agonist performance in Planaria after manganese treatment. Experientia. 36:449–50.

Palladini G et al. (1981) Photosensitization and the nervous system in the *Planaria Dugesia gonocephala*. A histochemical, ultrastructural and behavioural investigation. Cell Tissue Res. 215:271–79.

Palladini G et al. (1983) The cerebrum of *Dugesia gonocephala* s.l. (Platyhelmintes, Turbellaria, Tricladida). Morphological and functional observations. J. Hirnforsch. 23:165–72.

Palladini G et al. (1988) The answer of the planarian *Dugesia gonocephala* neurons to nerve growth factor. Cell Mol Biol. 34(1):53–63.

Palladini G et al. (1996) A pharmacological study of cocaine activity in planaria. Comp Biochem Physiol C Pharmacol Toxicol Endocrinol. 115(1):41–45.

Passarelli F et al. (1999) Opioid-dopamine interaction in planaria: a behavioral study. Comp Biochem Physiol C Pharmacol Toxicol Endocrinol. 124(1):51–55.

Pearl R (1903) The movements and reactions of fresh-water planarians: a study of animal behavior. Quart J Micr Sci. 46:509–714.

Pedersen KJ (1958) Morphogenetic activities during planarian regeneration as influenced by triethylene melamine. J Embryol Exp Morphol. 6(2):308–34.

Petersen CP, Reddien PW (2008) Smed-betacatenin-1 is required for anteroposterior blastema polarity in planarian regeneration. Science. 319(5861):327–30.

Pierce WD (1960) Silicified turbellaria from calico mountains nodules. Bull So Calif Acad Sci. 59(3):138–43.

Pigliucci M (2007) Do we need an extended evolutionary synthesis? Evolution. 61(12): 2743–49.

Poinar G (2003) A rhabdocoel turbellarian (*Platyhelminthes, Typhloplanoida*) in Baltic amber with a review of fossil and sub-fossil platyhelminths. Invertebrate Biol. 122(4):308–12.

Porfireva NA (1969) Endemichnaia Baikal'skaia Planariia *Rimacephalus arecepta* sp. n. (Tricladida, Paludicola) (An endemic Baikal planarian *Rimacephalus arecepta* sp. n. [Tricladida, Paludicola]). Zoologicheskii Zhurnal. 48:1303–8.

Powell K (2010) Masters of regeneration. HHMI Bull. 23(3):28–48.

Prestwich GD (1984) Interspecific variation of diterpene composition of Cubitermes soldier defense secretion. J Chem Ecol. 10:1219–31.

Rachamim T, Sher D (2012) What Hydra can teach us about chemical ecology - how a simple, soft organism survives in a hostile aqueous environment. Int J Dev Biol. 56(6–8):605–11.

Raffa RB et al. (2005) Bacterial communication ("quorum sensing") via ligands and receptors: a novel pharmacologic target for the design of antibiotic drugs. J Pharmacol Exp Ther. 312(2):417–23.

Raffa RB, Desai P (2005) Description and quantification of cocaine withdrawal signs in planaria. Brain Res. 1032(1–2):200–202.

Raffa RB, Valdez JM (2001) Cocaine withdrawal in Planaria. Eur J Pharmacol. 430(1):143–45.

Ramón y Cajal S (1906) Nobel lecture. http://www.nobelprize.org/nobel_prizes/medicine/laureates/1906/cajal-lecture.pdf

Randolph H (1897) Observations and experiments on regeneration in Planarians. Archiv für Entwicklungsmechanik der Organismen 5(2):352–72.

Raviola E, Mazzarello P (2011) The diffuse nervous network of Camillo Golgi: facts and fiction. Brain Res Rev. 66:75–82.

Rawls SM et al. (2008) Agmatine: identification and inhibition of methamphetamine, kappa opioid, and cannabinoid withdrawal in planarians. Synapse. 62(12):927–34.

Rawls SM et al. (2011) Nicotine behavioral pharmacology: clues from planarians. Drug Alcohol Depend. 118(2–3):274–79.

Reddien PW et al. (2008) Gene nomenclature guidelines for the planarian Schmidtea mediterranea. Dev Dyn. 237(11):3099–101.

Redfern WS et al. (2008) Zebrafish assays as early safety pharmacology screens: paradigm shift or red herring? J Pharmacol Toxicol Methods. 58(2):110–17.

Rehm H, Gradmann D (2010) Plant neurobiology, intelligent plants or stupid studies. Lab Times, March:30–32. http://www.labtimes.org/labtimes/issues/lt2010/lt03/lt_2010_03_30_33.pdf

Reichert H, Simeone A (2001) Developmental genetic evidence for a monophyletic origin of the bilaterian brain. Philos Trans R Soc Lond B Biol Sci. 356(1414):1533–44.

Reuter M, Gustafsson MK (1995) The flatworm nervous system: pattern and phylogeny. EXS. 72:25–59.

Ribeiro P et al. (2005) Classical transmitters and their receptors in flatworms. Parasitology 131 Suppl:S19-40.

Richter S et al. (2010) Invertebrate neurophylogeny: suggested terms and definitions for a neuroanatomical glossary. Front Zool. 7:29. doi:10.1186/1742-9994-7-29

Rieger RM (1998) 100 years of research on 'Turbellaria'. Hydrobiologia. 383:1–27.

Rilling M (1996) The mystery of the vanished citations: James McConnell's forgotten 1960s quest for planarian learning, a biochemical engram, and celebrity. Am Psychol 51(9):589–98.

Riutort M et al. (2012) Evolutionary history of the Tricladida and the Platyhelminthes: an up-to-date phylogenetic and systematic account. Int J Dev Biol. 56(1–3):5–17.

Rivera MC, Lake JA (2004) The ring of life provides evidence for a genome fusion origin of eukaryotes. Nature. 431(7005):152–55.

Rodríguez AD (1995) The natural products chemistry of West Indian gorgonian octocorals. Tetrahedron. 51(16):4571–18.

Rohde K (2000). Platyhelminthes (flatworms). eLS. John Wiley & Sons Ltd.http://www.els.net. doi:10.1038/npg.els.0001585

Rosner D, Markowitz G (2005) Standing up to the lead industry: an interview with Herbert Needleman. Public Health Rep. 120(3):330–37.

Rossi L et al., (2008) Planarians, a tale of stem cells. Cell Mol Life Sci. 65(1):16–23.

Rounak S (2011) Zoopharmacognosy (animal self-medication): a review. IJRAP. 2(5):510–12.

Rowlands AL, Pagán OR (2008) Parthenolide prevents the expression of cocaine-induced withdrawal behavior in planarians. Eur J Pharmacol. 583(1):170–72.

Rubin RP (2007) A brief history of great discoveries in pharmacology: in celebration of the centennial anniversary of the founding of the American Society of Pharmacology and Experimental Therapeutics. Pharmacol Rev. 59(4):289–59.

Ruetsch Y et al. (2001) From cocaine to ropivacaine: the history of local anesthetic drugs. Curr Top Med Chem. 1(3):175–82.

Saigusa T et al. (2008) Amoebae anticipate periodic events. Phys Rev Lett. 100(1):018101.

Saito Y et al. (2003) Mediolateral intercalation in planarians revealed by grafting experiments. Dev Dyn. 226(2):334–40.

Saló E (2006) The power of regeneration and the stem-cell kingdom: freshwater planarians Platyhelminthes). Bioessays. 28(5):546–59.

Saló E et al. (2009) Planarian regeneration: achievements and future directions after 20 years of research. Int J Dev Biol. 53(8–10):1317–27.

Saló E, Agata K (2012) Special issue: The planarian model system. International journal of developmental biology 56(1–3):1–196.

Saló E, Baguñà J (2002) Regeneration in planarians and other worms: New findings, new tools, and new perspectives. J Exp Zool. 292(6):528–39.

Saló E, Baguñà J (1984) Regeneration and pattern formation in planarians. I. The pattern of mitosis in anterior and posterior regeneration in *Dugesia (Girardia) tigrina*, and a new proposal for blastema formation. J Embryol Exp Morphol. 83:63–80.

Sánchez Alvarado A (2000) Regeneration in the metazoans: why does it happen? Bioessays. 22(6):578–90.

Sánchez Alvarado A (2012) Q&A: what is regeneration, and why look to planarians for answers? BMC Biol. 10:88.

Sarnat HB, Netsky MG (1985) The brain of the planarian as the ancestor of the human brain. Can J Neurol Sci. 12(4):296–302.

Sarnat HB, Netsky MG (2002) When does a ganglion become a brain? Evolutionary origin of the central nervous system. Semin Pediatr Neurol. 9(4):240–53.

Saxe CL (1999) Learning from the slime mold: Dictyostelium and human disease. Am J Hum Genet. 65(1):25–30.

Schulz S et al. (2003) New terpene hydrocarbons from the alligatoridae (crocodylia, reptilia). J Nat Prod. 66(1):34–38.

Schwarz D et al. (2011) Parthenolide blocks cocaine's effect on spontaneous firing activity of dopaminergic neurons in the ventral tegmental area. Curr Neuropharmacol. 9(1):17–20.

Shimizu H et al. (1993) Minimum tissue size required for hydra regeneration. Dev Biol. 155(2):287–96.

Shomrat T, Levin M (2013) An automated training paradigm reveals long-term memory in planaria and its persistence through head regeneration. J Exp Biol. 216(Pt 20):3799–810

Sikes JM, Newmark PA (2013) Restoration of anterior regeneration in a planarian with limited regenerative ability. Nature. 500(7460):77–80.

Silber RH, Hamburger V (1939) The production of duplicitas cruciata and multiple heads by regeneration in *Euplanaria tigrina*. Physiol Zool. 12(3):285–301.

Simon A (2013) Regenerative biology: On with their heads. Nature. 500(7460):32–3.

Sluys R et al. (2009) A new higher classification of planarian flatworms (Platyhelminthes, Tricladida). J Nat Hist. 43(29–30):1763–77.

Sluys R, Kawakatsu M, Bleeker J (2006). The Kawakatsu Collection incorporated within the collection of the Zoological Museum Amsterdam. Ber Nat-Med Verein Innsbruck. Suppl. 16:89.

Sporns O et al. (2005) The human connectome: a structural description of the human brain. PLoS Comput Biol. 1(4):e42.

Stahlberg R (2006) Historical overview on plant neurobiology. Plant Signal Behav. 1(1):6–8.

Stephan-Dubois F, Gilgenkrantz F (1961) Transplantation and regeneration in the planarian, *Dendrocoelum lacteum*. J Embryol Exp Morphol. 9:642–49.

Styx G (2005) A toxin against pain. Sci Am. 292(4):70–75.

Sullivan RJ, Hagen EH (2002) Psychotropic substance-seeking: evolutionary pathology or adaptation? Addiction 97:389–400.

Swanson LW (2000) What is the brain? Trends Neurosci. 23(11):519–27.

Tan TC et al. (2012) Telomere maintenance and telomerase activity are differentially regulated in asexual and sexual worms. Proc Natl Acad Sci U S A. 109(11):4209–14.

Tanaka EM, Reddien PW (2011) The cellular basis for animal regeneration. Dev Cell. 21(1):172–85.

Terlau H, Olivera BM (2004) *Conus* venoms: a rich source of novel ion channel-targeted peptides. Physiol Rev. 84(1):41–68.

Thompson R, McConnell J (1955) Classical conditioning in the planarian, Dugesia dorotocephala. J Comp Physiol Psychol. 48:65–68.

Torres GE et al. (2003) Plasma membrane monoamine transporters: structure, regulation and function. Nat Rev Neurosci. 4(1):13–25.

Twede VD et al. (2009) Neuroprotective and cardioprotective conopeptides: an emerging class of drug leads. Curr Opin Drug Discov Devel. 12(2):231–39.

Tyler S (1999) Systematics of the flatworms-Libbie Hyman's influence on current view of the Platyhelminthes. In: Libbie Henrietta Hyman: Life and Contributions. Edited by Winston JE. Novitates 3277:1–66.

Uhl G et al. (2002) Cocaine, reward, movement and monoamine transporters. Molec Psychiatry. 7:21–26.

Umesono Y et al. (2011) Regeneration in an evolutionarily primitive brain—the planarian Dugesia japonica model. Eur J Neurosci. 34(6):863–69.

Umesono Y, Agata K (2009) Evolution and regeneration of the planarian central nervous system. Dev Growth Differ. 51(3):185–95.

Umesono Y et al. (2013) The molecular logic for planarian regeneration along the anterior-posterior axis. Nature. 500(7460):73–6.

Valmalette JC et al. (2012) Light- induced electron transfer and ATP synthesis in a carotene synthesizing insect. Sci Rep. 2(579):1–5.

van Oye P (1920) Over het geheugen bij de Platwormen en andere Biologische waarnemingen bij deze dieren. Natuurwet Tijdschr. 2:1–9.

Van Regenmortel MH (2004a) Biological complexity emerges from the ashes of genetic reductionism. J Mol Recognit. 17(3):145–48.

Van Regenmortel MH (2004b) Reductionism and complexity in molecular biology. EMBO Rep. 5(11):1016–20.

Venturini G et al. (1981a) Naloxone enhances cAMP levels in Planaria. Comp Biochem Physiol C. 69:105–8.

Venturini G et al. (1981b) Ouabain insensitive ATPase in planaria. Comp Biochem Physiol B. 70:775–78.

Venturini G et al. (1983) Radioimmunological and immunocytochemical demonstration of met-enkephalin in planaria. Comp Biochem Physiol C. 74:23–25.

Venturini G et al. (1989) A pharmacological study of dopaminergic receptors in planaria. Neuropharmacology. 28(12):1377–82.

Vetter I, Lewis RJ (2012) Therapeutic potential of cone snail venom peptides (conopeptides). Curr Top Med Chem. 12(14):1546-52.

Vispo M et al. (1996) Regionalisation along the anteroposterior axis of the freshwater planarian *Dugesia (Girardia) tigrina* by TCEN49 protein. Int J Dev Biol. 1:209S–10S.

Wagner DE et al. (2011) Clonogenic neoblasts are pluripotent adult stem cells that underlie planarian regeneration. Science. 332(6031):811–16.

Wahlberg I, Eklund AM (1992) Cembranoids, pseudopteranoids and cubitanoids of natural occurrence. Prog Chem Org Nat Prod 59:141–294.

Walter HE (1908) The reactions of planaria to light. Exp Zool. 5:35–163.

Wang E et al. (2004) Transgenic *Nicotiana tabacum l.* with enhanced trichome exudate cembratrieneols has reduced aphid infestation in the field. Mol Breeding. 13(1):49–57.

Welch GR, Clegg JS (2010) From protoplasmic theory to cellular systems biology: a 150-year reflection Am J Physiol Cell Physiol. 298:C1280–90.

Wiemer DF et al. (1979) Cembrene A and (3Z)-cembrene A: diterpenes from a termite soldier (*Isoptera: Termitidae; Termitinae*). J Org Chem 44:3950–52.

Wigner E (1960) The unreasonable effectiveness of mathematics in the natural sciences. Commun Pure Appl Math. 13:1–14.

Williams P et al. (2000) Quorum sensing and the population-dependent control of virulence. Philos Trans R Soc Lond B Biol Sci. 355(1397):667–80.

Williams SM et al. (2004) The use of animal models in the study of complex disease: all else is never equal or why do so many human studies fail to replicate animal findings? BioEssays. 26:170–79.

Winston JE (1991) Libbie Henrietta Hyman: life and contributions. Novitates 3277:1–66.

Woese CR (2004) A new biology for a new century. Microbiol Mol Biol Rev. 68(2):173–86.

Index

Pages in *italic* indicate plates and illustrations.

A. *See* adenine (A)
"About Memory in Flatworms and Other Biological Observations in These Animals" (van Oye), 170
absolute zero, 80
acetylcholine, 54, 55, 57, 59, 179, 180
　nicotine mimicking, 88
　nicotinic acetylcholine receptor (nAChR), 85, 185, 186, 189, 190, 191, 211ch10n39
Ackerman, Diane, 62
acoels, *Plate V*
Acoels, 96, 99, 101, 160
acquired characteristics, inheritance of, 118, 119
action potential, 53, 54, 58
active site, 80
adenine (A), 22
adenosine triphosphate (ATP), 21, 31–2
adhesion, 8
ADME scheme (administration, distribution, metabolism, and excretion), 82
"The Adventures of Planarian Man" (Obermeyer), 140

affinity (interaction between compound and molecular target), 79, 81–2
Agata, Kiyokazu, 166, 180
"Age of Wonder," 118, 126
aging (senescence), 136
Alcmaeon, 69
alcohol, human reactions to, 32, 88, 90–1
algae, green, 127, 155
alkaloids, 83–4, *84*, 85, 89
allele, 119–20
Alligator sinensis, 188
alloeocoel turbellarian, *106*
ALS (motor neuron disease), 165, 207ch9n31
alternative splicing, 74
Alzheimer's disease, 77, 165, 186
amber, flatworm fossil in, *103*
Ambystoma maculatum, 155
American Museum of Natural History, 113
American Naturalist (journal), 178
amino acids, 22–3, *23*, 36
amoeba, 58, 152
amphetamines, 57, 176, 177, 192
amyotrophic lateral sclerosis. *See* ALS (motor neuron disease)

Index

anatomy, 11, 70, 157, 158, 160, 168, 176, 179
  simplified general anatomy of a typical triclad, 97
anesthesia, local, 85–7, 185, 190
  cocaine as an anesthetic, 85, 86, 87, 89, 178, 191, 192, 193
Angraecum sesquipedale, 13–14
animals, 25. *See also* invertebrates; specific names of species; vertebrates
  animal models in pharmacology, 90–2
  bilaterian animals, 96, 99, 100, 101, 131, 134
  chemicals used as weapons, 183–4
  distinguishing animals from plants, 127
  frequency of ability to regenerate, 133
  planarians as first hunters in animal kingdom, 107
  recognizing that some organisms not "better" than others, 27, 200ch2n7
  self-medication of, 78–9
  use of in research, 26–31, 90
    problem of genetic homogeneousness, 30, 32
    reasons for using planarians in, 176
    reproduction as criterion for choosing, 90, 122, 134
    use of planarians in, 164–8, 178–83, 180, *181*, 181–3, 192–4
Anolis genus, 125
anterior-posterior axis (A-P axis), 157
antibiotic resistance, 152
anti-inflammatory effects, 188–9
antitumor compounds, 188
ants, 188
A-P axis. *See* anterior-posterior axis (A-P axis)
applied science, 34
Arabidopsis thaliana, 29
archaea, 25, 26, 151
Aristotle, 14, 37, 62, 69, 125
aroma, 8, 188
artificial selection, 88
artist and scientist, 4
Asimov, Isaac, 143
association rate, 80–1, 82
ATP. *See* adenosine triphosphate (ATP)
ATPase, 51–2
atropine, 153
Auburn Press-Tribune (newspaper), 139, 140
auricles, 104, 105
autism, 186

autolyze (dissolving head first), 102
autoreceptors, 74–5, 81, 87
autotomy, 125
autotrophs, 21, 155
Axelrod, Julius, 202ch5n6
axon, 42, 43, 48, 50

Babylon 5 (TV show), 142
bacteria, 25, 26, 151–2, 194
Baguñà, Jaume, 135
baldscientist.wordpress.com (blog), xxv
"basal" bilaterians, 131, 134
basal ganglia, 62, 63
basic science vs. fundamental science, 33–7, 84
Battlestar Galactica (TV show), 141–5
  similarity of Cylon raider to *Duplicitas cruciata* worm, *146*
Bdellocephala annandalei, Plate I
behavior, 21, 28, 60, 78, 90, 143, 151, 156, 179, 194. *See also* learning; memory research
  addictive behavior, 54, 87, 159, 176, 177, 192
  bacterial social behavior, 151
  behavioral pharmacology, 91–2, 176, 180
  and the central nervous system/brain, 156–8, 168–9
  of cnidarians, 153
  and communication, 61, 152
  ethology (animal behavior), 169
  of hydra, 127
  impact of "second brain" on, 67
  and intelligence, 158
  of molecules, 143
  and neurons, 11, 42, 44, 76
  of planarians, 105–7, 111, 158, 168–74, 176, 177
    complex behaviors, 107, 176
    withdrawal-like behaviors, 177–8, *178*, 180, 182, 192
  and plants, 21, 58–9, 60, 61, 127
  of polyclads, 98
  predicting, 76
  protopsychology, 168–74
  as a reaction to environment, 58, 151
  without a nervous system, 59, 151, 152, 153–4
Benazzi, Mario, 134
Bennet, Michael, 173
Best, Jay B., 169

Bialik, Mayim, 147
The Big Bang Theory (TV show), 147
bilateral symmetry, 99, 100, *100*
bilaterian animals, 96, 99, 100, 101, 131, 134
bilobar brain, xiv, 63, 66, 161
binding, 167, 185
   binding pocket (active site), 80
   "everything starts with binding," 79, 167, 184
   interaction between compound and molecular target, 79, 80–1
   mutually exclusive binding, 190–1
binomial system of nomenclature, 25
biochemistry, 12, 19–31
   and the classification of life, 24–31
biodiversity, 37, 57
biofilm, 151
Biological Society of Kinki Normal Colleges, 115
biology
   biological research, 29–30, 90. *See also* animals, use of in research; biomedical research
      evaluating quality of organisms for research, 27, 200ch2n7
      need for interdisciplinary approach, 127
      practicality as a factor in research, 122
      reproduction as criterion for choosing animals, 90, 122, 134
      use of planarians in, 164–8, 176, 178–83, *180*, *181*, 181–3, 192–4
   biology of the nervous system. *See* neurobiology
   cell biology, 19–24. *See also* cells
   developmental biology, 123
   evolutionary biology, 14–18
   Lamarck giving modern meaning of, 118
   neurobiology, 12
   organismal biology, 9–10
   as a science of exceptions, 50, 100, 119
   structural biology, 11
   studying from a molecular perspective, 9, 10, 11, 12
biomedical research, 19–37, 34
   and animal models, 26–31
      problem of genetic homogeneousness, 30, 32
   basic science vs. fundamental science, 33–7
biophysics, 12
Bird, C., 60

bisepieupalmerin, *187*
Bishop, Peter (fictional character), 146
Bishop, Walter (fictional character), 146
Bitterman, M. E., 171, 172, 208ch9n44
"black reaction" (staining technique), 47, 48, 49
Blanco y Pagán, Ada, 115
blended inheritance theory, 118–19
blood type, 119–20
Boltzmann kinetic theory of gases, 143
Botox, 57
botulinum toxin, 57
Brabbinthes churkini, 103
brain, 62–77, *65*
   bilobar brain, xiv, 63, 66, 161
   building blocks of, 11–12
   and the cell theory, 45
   complexity of, 11–12
   development of early nervous systems without brains, 151–4
   protobrains, 160, *Plate V*
   difficulty in defining, 62–4, 156–7, 158
   first hunters, 154–6
   and generation of behavior, 156–8, 168–9
   hemispheres of the brain, 64–5
   human brain, 62–77, *65*, 165, 186, 194–5, 198
      ability to heal after damage to, 125, 131
      and addiction, 192, 193
      brain connections changing, 75
      comparison of planarian and human nervous system, 162, *163*
      as a complex system, 71–7
      computer simulation of, 76
      as control center of body, 67
      as a double organ, 64–5
      hardwiring in the brain, 74
      historical understanding of, 66–71, *68*
      mapping of, 72–3, 74–5, 77
      and the nervous system, 64–6. *See also* nervous system
      overview of, 62–4
      planarian brain teaching about human brain, 194–5
      plasticity of, 64, 125
      size of, 64, 71, 76, 201ch4n11, 202ch4n16
      theories of brain architecture, 62–4
   location of in relation to digestive system, 168

brain (*Cont.*)
  nerve cells in, 8–9, 11, 44. *See also* glial cells; neurons
  planarian and flatworm brains, xxiii, 97, 128, 137, 158–9, 160–2, 162, 165, 198
    ability to learn, 169–71, 173
    as the "ancestor" of vertebrate brains, 158, 159–64
    brain transplants in, 98
    comparison of planarian and human nervous system, 162, 163
    and FGFR-related gene, 208ch9n32
    regeneration of, 165–8, 172, 173, 176
    representation of typical brain, 162
    teaching about human brain, 194–5
    well-organized nervous system, 159–64
  ultrastructure of, 11
brainstem, 65
British Society for Developmental Biology, 135
Brønsted, Holger Valdemar, 109, 136, 145
Bryn Mawr College, 121
BSG. *See Battlestar Galactica* (TV show)
Burdon-Sanderson, John, 59
burial site, medicinal plants found at, 89
Buttarelli, Francesca, 179

C. *See* cytosine (C)
Caenorhabditis elegans, 29, 76, 100
caffeine, 84
Cajal, Ramón Y. *See* Ramón y Cajal, Santiago
calcium, 51, 56
Calvin, Melvin, 173
cancer, 32, 37, 112, 165, 188
carbohydrates, 19, 20
carbon in alkaloids, 84
carbon monoxide, 75
carnivores. *See also* predators
  animals as, 59, 107, 155–6
  carnivorous plants, 59, 127, 155
Carolei, Antonio, 179
cars, process of learning about, 10–11
catenulida, 96
Cebrià, Francesc, 165–6
cell biology, 19–24, 46, 168, 176
cells, 20, 30, 91, 131–2, 151, 152, 167. *See also* eukaryotes; prokaryotes; receptors; signaling
  all cells (except sex cells) in a given body having same genome, 131–2

and ATPase, 51–2
as building blocks, 11, 12
in *Caenorhabditis elegans*, 29
cell-based research, 30. *See also* ex vivo experiments
cell proliferation, 123. *See also* regeneration
cellular communications, 11, 20, 50, 55–6
  excitable cells, 50–3
  and cooperation, 58, 131, 151, 152
differentiation of, 132
excitable cells, 75
glial cells, 11, 75–6
interstitial cells, 132
nematocysts, 127
neuronal cells, 42, 44, 45, 46, 47, 51, 56, 59, 63, 67, 72, 157, 163, 201ch4n11. *See also* synapses
number of cells needed to allow for regeneration
  of hydra, 127–8, 129
  of planarians/flatworms, 130, 205ch7nn23,24
pluripotent cells, 132
and signaling, 55–6, 58, 80, 152–3
somatic cells, 76
of squid, 29
stem cells, 129, 132, 133
subcellular machinery, 59
cell theory, 45–6, 118
cembranetrienediol, 187
cembranoids, 187–9, 191, 193
  cembrene ring, 187
  comparison of parthenolide with euniolide, 190
  as nicotine antagonists, 189, 192
  representative cembranoid-like molecules, 187
centralization, 157
central nervous system (CNS), 64, 65, 159, 163. *See also* nervous system
cephalic ganglia, *Plate VII*
cephalization, 101, 157, 160
cephalochordates, xiv
cerebellum, 62, 65–6
cerebral cortex, 62, 65–6
cerebrum, 65–6
Cestoda, 96
Chamovitz, Daniel, 61
cheetah, 31, 99
chemical ecology, 83, 184

chemical means of communication, 59, 153, 184
chemical neurotransmission, 53–7
chemical synapses, 53–7, 54, 55, 87
chemistry, 5, 19–20. *See also* biochemistry
Child, Charles Manning, 113, 121, 123, 124, 131, 206ch8n10
chimpanzees, 32
Chinese alligator (*Alligator sinensis*), 188
chiral nature of amino acids, 22
chlorophyll, 84, 155
chloroplasts, 155
chordates, 164
cilia, 97
classification system for life, 24–31, 95, 99
  correct classification of a flatworm, 104, 160
  misclassification of planarians, 103
"Classsical Conditioning in the Planarian, *Dugesia dorotocephala*" (Thompson and McConnell), 172
cnidarians, 99, 153–4
Cnidaria phylum, 126
cnidocysts, 106
CNS. *See* central nervous system (CNS)
Cobo, Bernabé, 86, 89
cocaine, xi, 57, 84, 85–8, 153
  as an anesthetic, 85, 86, 87, 89, 178, 191, 192, 193
  planarians reacting to, 178, 179, 207ch9n23
  proteins as targets of abused drugs, 181
"Cocaine Withdrawal in Planarians" (Raffa and Valdez), 180
coca leaves, 86
codominance, 119–20
codons, 22
coelacanth, 26
coffee, flavor of, 8
cognition, 60, 65, 144, 158
cohesion, 8
coiling reflex, xiv
collectives, slime molds creating, 152
Columbia University, 121
comic books and planarians, 139–41, 142
common ancestor, 18
communication, 33, 58
  and behavior, 61, 152
  between cells, 11, 20, 50, 55–6, 151, 152–3, 168
    excitable cells, 50–3

chemical means of, 59, 153, 184
and ion channels, 185
within the nervous system, 42, 44, 50–1, 61, 63, 72, 74
in simpler organisms, 150–1
Comparative Vertebrate Anatomy (Hyman), 113
complex systems, 6–7, 7, 10–12, 200ch2n7
  dealing with complex environments, 194
  and emergence, 7–9
computer modeling research in biology, 30
computer simulation of the human brain, 76
concentration-dependent responses to drugs, 177
cone snails, 35–7
connectome
  of *Caenorhabditis elegans*, 76
  and the human brain, 72–3, 74
  brain activity map, 75, 76–7, 77
conotoxins, 36–7, 57
consciousness, and emergent properties, 8–9
conservation of resources, need for, 209ch10n1
conservation of traits, 164
cooperation, 35, 74, 156
  between cells, 58, 131, 151, 152
corals, 126, 184, 188
corkscrew, 178
Cornell University, 181–2, 210ch10n18
Correns, Carl, 120
cotransmission, 75
Courtney-Latimer, Marjorie, 26
cranial nerves, 66, 162
crassin acetate, 187
Crawford, Mary, 112
Crick, Francis, 44, 151, 183
curiosity, xii, xxii–xxiii, 5, 57, 141–2, 183, 197, 198
currents, sensitivity to, 106
Cylons (fictional characters), 144
cytosine (C), 22

Daily Nebraskan (newspaper), 139
Dalton (unit of measure), 184
Dalyell, John Graham, 108, 110, 110, 111, 144
Danio rerio, 28–9
Daphnia, 107
Darwin, Charles, 13, 14–15, 59, 88, 118, 119
  visit to Hobart, Tasmania, 197, 212n1
Darwin, Emma, 59
Darwin Orchid (Comet Orchid), 13–14

DAT. *See* dopamine
decussating interneurone, xiv
defense, chemicals used as, 184
dendrites, 42, 43, 48, 59
   dendritic spines, 161
Dendrocoelopsis hymanae, 114
Dendrocoelum lacteum, 98
dendrology, 59
deoxyribose, 21
depolarization, 53
Descartes, René, 6–12
desensitization, 56
developmental biology, 123, 164
developmental origin of the brain, 62
DeVries, Hugo, 120
differentiation, 132
diffusion, 56
digestive system, 51, 67, 168
   of cnidarians, 127
   of hydra, 127
   of planarians, 97. *See also* pharynx
   as a "second brain," 67
Dilepti, 106–7
dimethylsulphoxide (DMSO), 191–2
Dionaea muscipula, 59
direction, sense of, 105–6
directionality of nerve transmissions, 48, 50
directional signaling, 153
dissociation, 10, 81, 82
   and regeneration, 125, 127, 205ch7n23
distribution of drugs, 82
DMSO. *See* dimethylsulphoxide (DMSO)
DNA, 9, 17, 21–2, 23, 24, 25, 27, 123
   epigenetics, 119
Dr. Planarian (Planarian Man's arch enemy), 140
Dr. Who (TV show), 147
Domains (as category in classification), 24–5
dominance, incomplete, 119
dopamine, 55, 179, 180
   and cocaine, xi, 87, 88
   dopamine transporter (DAT), 54, 87, 182, 191
   extrasynaptic dopamine autoreceptor, 87
dopaminergic system of planaria and cocaine, 179
dorsal spinal cord, 159, 161
Drosophila melanogaster, 28, 121, 122, 123, 124, 129

drugs, 78. *See also* pharmacology
   abuse of, 57, 78, 91, 107, 159, 193. *See also* specific drugs, i.e., cocaine, nicotine, etc.,
   alkaloids of, 84, *84*
   cure for addiction not yet available, 193
   and dopamine, 54–5
   and fruit flies, 28
   and planarians, 147, 169, 175–8, 181, 192, 193–4
   proteins as targets of abused drugs, 181
   as secondary alkaloids, 84–5
   and sponges, 153
drug metabolism, 82, 91
   harmful effects of, 79
   interactions between compound and molecular target, 79–82
   path a drug takes in the body, 82
   study of drug targets and ligands, 82
dual brain, 62
Dubois, François, 131
Dugesia austroasiatica, Plate III
Dugesia dorotocephala, 179
Dugesia japonica [aka Dugesia gonocephala], 98, 115, 135, 136, 161, 166, 168–74, 175, 179, Plate III
   brain and central nervous system of, Plate VIII
   central nervous system of, Plate VI
Dugesia lugubris [aka Dugesia mediterranea]. *See Schmidtea mediterranea*
Dugesia ryukyuensis, Plate III
Dugesia tigrina [aka Dugesia maculata]. *See Girardia tigrina*
Dunham, Olivia (fictional character), 146
duplicitas cruciata, 145
   and the *Battlestar Galactica* Cylon raider, *146*
dwarfism, 165
dynamic polarization theory, 48

earthworms, 69, 97, 111, 116, 125
echinoderms, xiv, 100
ecology, chemical, 83, 184
Edison, Thomas A., 41
Edwin Smith Papyrus, 68
EEG. *See* electroencephalogram (EEG)
Egyptian understanding of the brain, 68, *68*, 70

Ehlers, Ernst, 95–6
Ehrenberg, C. G., 97
Einstein, Albert, 3, 14
electrical synapses, 53, 153
electric eel, freshwater, 186
electric organs in fish, 186
electroencephalogram (EEG), 164
Electrophorus electricus, 186
electrophysiology, 50–3, 58, 59, 60, 61, 211ch10n39
  plant electrophysiology, 59–61, 155
elephant's trunk, 110
Elrich, Paul, 78, 79
Elysia chlorotica, 155
embryology, 123
emergent property, 7–9, 9, 199ch1n5
encephalon, 62
endorphins, 88
energy, capture of, 20–1
environment
  behavior as response to, 58
  the brain and complex environments, 194
  need for protection of, 209ch10n1
Eocene period, 102
epigenetics, 119
epilepsy, 186
epimorphosis, 123
equilibrium state, 81
Erasistratus, 70
Erythroxylum, 86
Eterovic, V. A., 189
ethology (animal behavior). *See* behavior
eukaryotes, 20, 25, 26, 104, 152. *See also* cells
euniolide, 187, 190
evolutionary biology, 9, 12–13
  energy needed for survival, 155
  evolutionary history of the brain, 63
  evolution as a Theory (logical model) rather than a guess, 14–18
  evolution not having directionality, 200ch2n7
  and genetics, 118–21
  and the human nervous system, 64
  new evolutionary synthesis (new synthesis) [aka neodarwinism], 120–1
  possible scheme of evolution of central nervous system, Plate IV
  reasons why plants create psychoactive substances, 82–3, 85, 88–9
  reproduction as the main goal of evolution and survival, 16, 21, 42, 90, 168, 198
  validation for, 58
"exaggerated" traits, 122
excitable cells, 50–3
excitatory neurotransmission, 56
excretion of drugs, 82
extremophiles, 25, 122
ex vivo experiments, 30, 91
eyes
  in planarians (ocelli), 104, 105, 130, 161
  sensitivity to/perception of light, 105, 139, 172, 206ch8n2
  usefulness of half an eye, 58

Fasciola punctata, 108
Ferchmin, P. A., 189
feverfew plant, 190
FGFR-related gene (fibroblast growth factor receptor), 165, 167, 208ch9n32
fight-or-flight response, 67
fish-hunting snails, 35–7
flatworms, xxiv, 92, 96, 106, 116, 131, 175–6, 179. *See also* planarians; *Platyhelminthes*; polyclads; triclads
  author's research on, 181–3
  biggest freshwater flatworm, 98
  as first hunting organism, 107, 158
  fossil records of, 101–2, *103*
  genome of, 135
  and marijuana, 147
  nervous system of, 160, 163, Plate V. *See also* brain, planarian
  overview on, 95–101
  parasitic flatworms, 96–7, 108
Foundation novels (Asimov), 143
Four O'clock (plant), 119
Freud, Sigmund, 86–7
Fringe (TV show), 145–6
fruit flies, 28, 121, 122, 123, 124, 129
fruiting body, 152
fundamental science vs. basic science, 33–7
fungi, 25, 35, 57
"Further Observations on Planariae" (Johnson), 111
The Future of Life (Wilson), 209ch10n1

G. *See* guanine (G)
GABA. *See* gamma-aminobutyric acid (GABA)

Galileo Galilei, 5, 209ch9n45
gamma-aminobutyric acid (GABA), 153
ganglia, 63, 163
   basal ganglia, 62, 63
gap junctions, 50
Gazzetta Medica Italiana, Lombardia (journal), 47
genetic drift, 15
genetics, 14–18. *See also* inheritance
   codominance, 119–20
   and the evolutionary process, 118–21
   and fitness, 17
   genetic differences, 16
   importance of genetic variability, 30, 31–3
   incomplete dominance, 119
   Mendelian genetics, 120–1, 123
   use of planarians to study, 121–5
genomes, 11, 15, 120, 131
   all cells (except sex cells) in a given body having same genome, 131–2
   human genome, 74, 90
   planarians and genomes, 132–6, 175
genomics, 90
genotype, 30–1, 123
genus (as category in classification), 25
germacranolides, 190
GiGo (garbage in, garbage out), 30
Girardia dorotocephala, 99, Plate II
Girardia tigrina, 99, 135, 169, 179, Plate III
glial cells, 11, 75–6
glutamate (glutamic acid), 54, 55, 59, 153
Golgi, Camillo, 46–50, 121
Gonzalez-Bonet, Gabriela, 211ch10n39
gorillas, 32
gossip, 44, 124, 183
Gould, Stephen Jay, 16, 199ch1n12
graphic arts and planarians, 137, 139–41
gravity
   sensitivity to, 106
   theory of, 14
Greek understanding of the brain, 62, 69, 70
green algae, 127, 155
guanine (G), 22
gyri, 65–6

habituation, 170
Hagmann, Patric, 72
Hallez, Paul, 97
Hamburger, V., 145
Hann, R. M., 189
haptors, 96, 102

The Harvard Law of Animal Behavior, 151
Hawking, Stephen, 207ch9n31
head
   anterior-posterior axis (A-P axis), 157
   formation of, 101
   head-specific genes in planarians, 166, 167
   regeneration of, 128, 204ch7n18
   representative morphology of, 106
headbop, 178
headswing, 178
healing as a form of regeneration, 125
heart as control center of body, 67
hemichordates, xiv
hemispheres of the brain, 64–5
Henry the Planarian, *xxvii*
herbivores, 155, 156, 158
heredity. *See* inheritance
hermit crabs, 153–4
Herophilus, 70
Hess, George Paul, 181–2, 192, 210ch10nn18,19
heterocyclic rings, 83–4
heterotrophs, 21, 155
Hippocrates, 70
history, 44
HIV. *See* human immunodeficiency virus (HIV)
Hobart, Tasmania, 197
holistic approach to studying nature, 46
homogenous genetic populations, 30, 31–3
Howard Hughes Medical Institute, 200ch2n13
human immunodeficiency virus (HIV), 32
humor and planarians, 138–9, *148*
Huntington's disease, 165
hybridization, 120
hydra, 100, 113, 126–8, 129, 132, 134
   diffuse nervous system of, 153, *154*
hydrophilic (water loving), 51
hydrophobic (water fearing), 51, 191
Hyman, Libbie Henrietta, 112–14, *117*, 124
hypothalamus, 62

Ichikawa, K., 175
immortality, 108, *110*, 144, 146, 179
immune system, 31
immunology, 98, 123
inbreeding, 31
incomplete dominance, 119
inflammation, physiological response to, 188–9

inheritance, 118–19, 120, 121, 123. See also genetics
inhibitory neurotransmission, 56
in silico experiments, 30
intelligence, 156, 158, 198
  in plants, 60, 61
interstitial cells, 132
invertebrates, 95, 160–1, 164, 168, 177. See also specific names of species, i.e., fruit flies, planarians, squids, etc.,
  and cembranoids, 188
  and kleptoplasty, 155
  not fossilizing well, 101–2
  use of in research, 29, 44, 85, 134, 159, 163–4
The Invertebrates (series by Hyman), 113–14
in vitro experiments, 30, 90
in vivo experiments, 28, 90
ion channels, 11, 36, 37, 53, 56, 80, 85, 185
ionotropic receptors, 56

jellyfish, 99–100, 126, 153–4
JEZ. See Journal of Experimental Zoology (JEZ)
Jiménez-Rivera, Carlos, 192
Johnson, J. R., 110–11
Johnson, Samuel, 64
Journal des Sçavans, 110
Journal of Comparative and Physiological Psychology, 172
Journal of Experimental Zoology (JEZ), 124
"Journey's End" (episode in *Dr. Who* TV show), 147

Kawakatsu, Masaharu, 114–15, *117*, 135, 175
Kenk, Roman, 115–17, *117*, 170, 204ch6n34
Kenk, Vida, 116–17, 204ch6n34
Kenkia, 116
keys and keyholes, 80
khoka, 86
Kimberella, 101
Kingdom (as category in classification), 25
kleptoplasty, 155
Koch, Cristof, 199ch1n5
Koller, Carl, 87
Komai, Taku, 114
Krebs cycle, 31–2
Krugman, Paul, 143, 206ch8n6
Kyôto Normal College (Kyôto Kyôiku University), 115, 180

A Laboratory Manual for Comparative Vertebrate Anatomy (Hyman), 113
A Laboratory Manual for Elementary Zoology (Hyman), 113
lagartijos, 125
Lamarck, Jean-Baptiste, 118, 119
Langley, John, 79
Latimeeria chalumnae, 26
lead exposure, toxicity of, 209ch10n4
learning, 28, 152, 153, 186. See also memory
  and planarians, 161, 169, 170, 171, 172–3, 174, 208ch9n44
  protopsychology, 168–74
  and survival, 57, 169
Leeuwenhoek, Antonie van, 47
Lent, Roberto, 201ch4n11
Levin, Mike, 174
Library of Congress, 116
lidocaine, 193
life, 6, 32, 42, 57–8, 90, 100, 132, 188, 194–5, 200ch2n7, 209ch10n1. See also amino acids; evolutionary biology; genetics; proteins
  based on DNA and RNA, 21–2. See also DNA; RNA
  biological definition of, 19, 20, 21
  border between life and nonlife, 23
  cells as basic unit of life, 11–12, 45, 51. See also cells
  characteristics of, 19–21
  as chemistry, 5, 12, 19, 27. See also biochemistry
  classification system for, 24–31, 95, 99
    correct classification of a flatworm, *104*, 160
    misclassification of planarians, 103
  complexity of, 34
  as an emergent property, 9, 12. See also emergent property
  energy currency of life, 52. See also ATPase
  and evolution, 14–18. See also evolutionary biology
  and kleptoplasty, 155
  life domains, 24–5, *26*
  and molecules, 9, 19
  oldest types of life on Earth, 151
  relationship of all life on earth, 17–18
  and science fiction, 143–4
  and symbiosis. See animals, 185–6
  unity of, 19–20, 26–7

life sciences. *See* natural science
ligands, 80–2
    ligand-gated ion channels, 85
light, sensitivity to/perception of, 105, 139, 172, 206ch8n2
Lincoln Journal-Star (newspaper), 140
Linné, Carl von (aka Linnaeus), 25
lipids, 19, 20, 51
locomotion, 96, 97, 111, 127
lophotoxin, 184–7, *185*, 189
LUCA (last universal common ancestor), 18, 26
luminescent bacteria, 151
lure, chemicals used as, 184

macromolecules, 12, 21, 22, 51, 83, 184. *See also* DNA; proteins; RNA
macrostomida, 96, Plate V
Macrostomum lignano, 97, 130–1, 135, 175, Plate III
Macrostomum pusillum, 131
macroturbellarians, 97
maps and the brain, 72–3, 74–5
    brain activity map, 76–7
Margotta, Vito, 179
Maricola (group), 97
marijuana and planarians, 147
marine ray, 186
Massachusetts Institute of Technology (MIT), 132
MAT. *See* monoamine transporter (MAT)
mathematics, 4, 5–6, 33
    and psychohistory, 143
    visualizing large numbers, 71–2
Mayr, Ernst, 9
McConnell, James V., 95, 146, 171–4, 208ch9n44
measurements and visualizing large numbers, 71–2
mechanistic explanation of evolution, 15–16
medulla, 62
membranes, 20
memory research, 161, 171, 208ch9n44
    chemical transfer of memory, 173
    and planarians, protopsychology, 168–74
    trained planarians fed to untrained planarians, 173
Mendel, Johann (Gregor), 120, 123
metabolism of drugs, 82, 91
metabolites, 84, 188

metabotropic receptors, 55
metazoans, 125, 153
microscope, invention of, 5–6
microturbellarian, 97, 130–1, Plate III
mimosa sp., 58
Miocene period, 102
MIT. *See* Massachusetts Institute of Technology (MIT)
Mlodinow, Leonard, xxi, xxii
modulators (neuromodulators), 75
molecular biology, 9–10, 11–12, 22, 61, 90, 120–1, 133, 168
molecules, 9. *See also* proteins
    and absolute zero (absence of molecular movement), 80
    association and dissociation, 81. *See also* ligands
    Boltzmann kinetic theory of gases, 143
    cembranoids, 187, 187–9
    charged molecules, 51
    interacting with other molecules, 190
    ion channels, 56
    and life, 9, 19
    macromolecules, 12, 21, 22, 51, 83, 184. *See also* DNA; proteins; RNA
    needing to look at whole organism, 159, 207ch9n23
    neurotransmitter molecules, 54, 56–7, 58, 74, 76, 87, 88, 161, 185. *See also* neurotransmission; signaling
    as receptors, 20. *See also* receptors
    and regeneration, 123, 133, 134
    small molecules, 183, 184
    and synaptic vesicles, 54
    toxins and venoms, 27, 35, 83
    lophotoxin molecules, 184, 189
mollusks, planarians classified as, 103
"Momentum Deferred" (episode in *Fringe* TV show), 146
monoamine transporter (MAT), 87
Monogenea, 96, 102
monomers, 21, 22
Morgan, Thomas Hunt, 121, 123, 124, 129–30, 131, 134, 206ch8n10
morphallaxis, 123
morphine, 86, 88, 176
morphology, 11, 16, 24, 100, 158, 168
    of planarians, 103, *106*, 111, 179
moths, 13–14

motility, 58, 177
motor neuron disease. *See* ALS
mouse ear cress, 29
movies and planarians, 146
Müller, Othone Friderico, 103, *105*, 108
mutations, 32, 90, 91, 100–1, 135. *See also* evolutionary biology; genetics
mutually exclusive binding, 190–1

nAChR. *See* nicotinic acetylcholine receptor (nAChR)
National Academy of Sciences, 210ch10n18
National Museum of Natural History (Smithsonian Institution), 116
natural science. *See also* biology
    development of, 3–6
    reductionist vs. holistic approach, 46
natural selection, 14–15, 16–17, 88, 118, 120. *See also* evolutionary biology; genetics; survival of the fittest
Nature (journal), 165
nature as a continuum, 3, 6
ndk. *See* nou-darake gene
Neanderthals and use of medicinal plants, 89
Needleman, Herbert L., 176, 209ch10n4
negative phototaxis, 105
nematocysts, 127
Nemertodermatids, 99, 101
neoblast theory, 131, 132, 167
neodarwinism, 120–1
nervous system, xiv, 63, 181. *See also* neurobiology; neurosciences
    acetylcholine, 54
    and behavior, 60, 156, 168, 194
        not needed to have behavior, 59, 151
    and the cell theory, 45
    central nervous system (CNS), 63, 64, 153–4, 156–7, 159, 160, 161, 163, 166. *See also* nuclei
        comparison of central nervous systems of 3 planarians, *Plate VII*
        examples of central nervous system arrangements, *Plate V*
        lack of, 64, 153
        possible scheme of evolution of central nervous system, *Plate IV*
    in cnidarians, 153–4
    development of, 57–9
    directionality of nerve transmissions, 48, 50. *See also* neurotransmission
    in fish, 36
    in flatworms, 98, 107, 128, 137, 159–60, 161–4, *162*, *163*, 167–8, 176, 179–80, 198
        brain and central nervous system of *Dugesia japonica*, *Plate VIII*
        central nervous system of *Dugesia japonica*, *Plate VI*
        comparison of central nervous systems of 3 planarians, *Plate VII*
    in humans, 8, 161, 194, 198
        and the brain, 64–6, *65*, *66*, 71–7
    in hydra, 127–8, 154
    in insects, 85
    ion channels, 56
    nerve cords and digestive system, relative arrangement of, 168
    nervous system-related genes, 164
    neurons as building blocks of, 42, 44, 63. *See also* neurons
    peripheral nervous system (PNS), 64, *65*, 159
    and plants, 83
    regenerative capacity of, 59, 125, 128, 131, 165, 167. *See also* regeneration
    in simpler organisms, 151–4
    in sponges (pseudo-nervous system), 153
    in squids, 29, 44
    and survival, 57–9, 194
    theories about composition of, 45–50
    use of dyes to make visible under microscope, 47, 48, 49
NET, 87
Netsky, Martin G., 64, 158
neurobiology, 12, 41, 59–61
neuroid systems, 153
neuromodulators, 75
neuronal sodium channels, 191
neurons, 42–4, *43*, 152, 157, 160. *See also* dendrites; nervous system
    as building blocks of the nervous system, 11–12
    process of establishing, 44–50
    in *Caenorhabditis elegans*, 29, 76
    in cnidarians, 153
    and communication, 50–3, 61, 63. *See also* signaling
    and dopamine transporters, xi, 182

neurons (*Cont.*)
    in flatworms, 161, 163, 177. *See also* brain, planarian
    in humans, 44, 186
        numbers of neurons, 71–7, 201ch4n11, 202ch4n16
    impact of cocaine on, 192
    and ion channels, 56, 185
    motor neuron disease. *See* ALS (motor neuron disease)
    neuronal protein, 192
    "pacemaker" neurons, 153
    plants not having neurons, 59, 60
    point of contact between neurons, 53–7, 54. *See also* neurotransmission; synapses
    ratio of glia-to-neurons, 75. *See also* glial cells
    shapes of, *43*
    similarity between species, 42, 44, 160
    in squids, 122
neuron theory, 44–50, 118
neuropeptides, 180
neuropharmacology, 176
Neuroscience Research Program Organization, 173
neurosciences, 41–61
    choice of planarians as animal model in neuroscience, 164–8
    interdisciplinary nature of, 41
    pyschology as behavioral neuroscience, 170–1
Neurosciences Research Program Bulletin (journal), 173
neurotransmission, 80, 153. *See also* acetylcholine; dopamine; gamma-aminobutyric acid (GABA); glutamate (glutamic acid); neuropeptides; nitric oxide; signaling
    atypical neurotransmitters, 75
    and autoreceptors, 74–5
    binding of transmitters to the receptor, 56
    chemical neurotransmission, 53–7, 55, 161
    cotransmission, 75
    and drugs, 57, 88, 180, 182, 191
    electrical neurotransmissions, 153
    and feedback mechanism, 50, 75
    first neurotransmitter discovered, 54
    and glial cells, 76
    impact of alkaloids on, 88
    interacting with proteins, 80. *See also* receptors
    and ion channels, 36, 37, 80
    neuromodulators, 75
    neurotransmitter spillover, 74
    neurotrasmitter transporters, 56
    in paramecium, 58
    in planarians, 161, 166, 180
    in plants, 59, 88
    reuptake of neurotransmitters, 87
    in sponges, 153
    terminating neurotransmissions, 56–7, 87
new evolutionary synthesis (new synthesis), 120–1
Newmark, Phil, 133, 135, 136, 166, 205ch7n31
Newton, Isaac, xxi, 14
nicotine, 84, *84*–5, 88
    cembranoids in tobacco plants, 189
    planarians reacting to, 175–8, 207ch9n23
    "relaxing" effect of, 177
nicotinic acetylcholine receptor (nAChR), 85, 185, 186, 189, 190, 191, 211ch10n39
Niemann, Albert, 86
nitric oxide, 75, 153, 180
nitrogen in alkaloids, 83–4
Nobel Prize in Physiology and Medicine, 48
Noble, Donna (fictional character), 147
nonneuronal cells. *See* glial cells
norepinephrine transporter (NET), 87
nou-darake gene (ndk), 165, 166–7, 208ch9n32
novocaine, 193
nuclei, 63
nucleic acids, 19
nucleotides, 21
numbers, visualizing large, 71–2

Obermeyer, Neal, 139–41
"Observations and Experiments on Regeneration in Planarians" (Randolph), 121
Observations of Some Interesting Phenomena in Animal Physiology, Exhibited by Several Species of Planariae (Dalyell), 108, 110, *110*
"Observations on the Genus Planarians" (Johnson), 111
ocelli (planarian eyes), 104, 105, 130, 161
octopi, 63, 103, 125, 168
Okajima, Ei'ichi, 114

Okugawa, K. I., 115
Olivera, Baldomero, 36
Omaha City Weekly (newspaper), 140
omnivores, 156
"On a New Organic Base in the Coca Leaves" (Niemann), 86
"On the Ecology and Distribution of Freshwater Planarians in the Japanese Islands, with Special References to Their Vertical Distribution" (Kawakatsu), 115
"On the Minimal Size of a Planarian Capable of Regeneration" (Montgomery and Coward), 130
Oophila amblystomatis, 155
open system, 20
ophthalmological anesthetic, 86–7
opiates, 88, 176, 177
optic nerves. *See* eyes
orchids, 13–14
organismal biology, 9–10
organisms. *See* animals; invertebrates; names of specific species, i.e., bacteria, flatworms, hydra, etc.,; vertebrates
orthogonal schemes, 160, 162

Palladini, Guido, 179–80
Pallas, Peter Simon, 108, *109*
Paludicola (group), 97
paramecium, 58
parasites, 129, 200ch2n7
parasitic flatworms, 96–7, 108, 129
parencephalon, 62
Parkinson's disease, 54, 165, 179
parthenolide, *190*, 190–2, 191, 193, 211ch10n39
Pasteur, Louis, 183
PCP. *See* phencyclidine (PCP)
pea plants, 120
peer review, 34
peptides (polypeptides), 23, *23*, 35, 36
neuropeptides, 180
peptide bond, 22, 23
perception of flavor and wetness, 8
Peripatus, 95
peripheral nervous system (PNS), 64, 65, 159
Permian period, 102
Perroncito, Aldo, 47
Phagocata kawakatsui, Plate I
Phagocata papillifera, Plate I, Plate II

Phagocata suginoi, Plate I
Phagocata ullala, Plate VII
Phagocata vivida, Plate I
pharmacodynamics, 82
pharmacogenetics, 91
pharmacogenomics, 91
pharmacokinetics, 82
pharmacology, xxiii, xxiv, 3, 12, 78–92, 183–4
    and animals, 90–2
        applying nonhuman models to humans, 192–4
        planarians, 137, 153, 158, 159, 164, 175–95
            behavioral pharmacology, 91–2, 176, 180
    chemicals used as weapons by living organisms, 183–4
    and the concept of affinity, 81–2
    and the concept of binding, 80–1
    and the concept of receptors, 79–81
    and discovery of new drugs, 34–5, 37
    impact of genetic homogeneity on, 32
    interdisciplinary nature of, 79
    and nicotinic acetylcholine receptor (nAChR), 185–6
    pioneers in the field, 79, 209ch10n4, 210ch10n18
    and plants, 82–9
    psychopharmacology, 82–9
    research in
        use of animal models, 27, 32, 112
        use of flatworms, 175–83, 180, *181*, 181–3, 191, 194
    as toxicology's sister science, 180. *See also* toxicology
    withdrawal-like behaviors, 177–8, *178*
pharynx, 103, 107, 110, 111, 163, 167, Plate VII
phases, 51
phencyclidine (PCP), *84*
phenotype, 31, 119, 123, 167
pheromone-like actions, 188
Philosophical Transactions of the Royal Society (journal), 110–11
phosphate group, 21–2
photoreceptors, 105
photosynthesis, 155
physical chemistry research, 34
physics, 5
    research, 34

Physiological Zoology (journal), 124, 145, 206ch8n10
physiology, 3, 10, 11, 16, 24, 27, 44, 48, 79, 82, 90–1, 158, 160, 207ch9n23
   of the brain, 62, 63, 65, 73, 76, 158–9
   of *Caenorhabditis elegans*, 29
   electrophysiology, 50–3, 58, 59, 60, 61, 210ch10n39
   of flatworms and planarians, 159, 160, 163, 164, 168, 175–6
   human physiology, 26–7, 28, 32, 33, 74, 85, 193
   Krebs cycle, 31–2
   neurotransmitters and synaptic physiology, 53, 54, 56, 57, 74, 75, 80, 88, 151, 185
   plant physiology, 60–1, 155
   and proteins, 22, 83
Pikaia, xiv
Planarian Man (superhero), 139–41, *142*
Planarian Regeneration (Brønsted), *109*
planarians, 95–117, *96*, 118–36
   ability to exist out of water, 107
   author's first discovery of, xxvi
   and behavior, 105–7, 111, 137, 158, 168–74, 176, 177
      complex behaviors, 107, 176
      withdrawal-like behaviors, 177–8, *178*, 180, 182, 192
   brain of, xiv, xxiii, 97, 128, 137, 158–9, 160–2, *161*, *162*, 165, 193–4, 198
      ability to learn, 169–71, 173
      as the "ancestor" of vertebrate brains, 158, 159–64
      brain transplants in, 98
      comparison of planarian and human nervous system, *162*, *163*
      and FGFR-related gene, 208ch9n32
      general structure of brain and nervous system, Plate VIII
      question of cognition in, 158, 207ch9n21
      regeneration of, 165–8, 172, 173, 176
      representation of typical brain, *162*
      teaching about human brain, 194–5
      well-organized nervous system, 159–64
   digestive system, 97
      pharynx of, *103*, 107, 110, 111, 163, 167, Plate VII
   double-headed planarians, 111
   and drug abuse, 147, 169, 175–8, 181, 192, 193–4
   early scientific work with, 108–11
   eyes (ocelli), 104, 105, 130, 161
   first use of term, 105
   and genomes, 132–6
   head-specific genes in, 166, 167
   Henry the Planarian, xxvii
   and humor, 138–9, *148*
   and learning, 161, 169, 170, 171, 172–3, 174, 208ch9n44
   and marijuana, 147
   and memory
      protopsychology, 168–74
      trained planarians fed to untrained planarians, 173
   misclassification of planarians, 103
   morphology of, *103*, *106*, 111, 179
   nervous system of, xiv, 161–4
      comparison of central nervous systems of 3 planarians, Plate VII
      comparison of human and planarian central nervous system, *163*
      comparison of planarian and human cranial nerves, *162*
      general structure of brain and nervous system, Plate VIII
      possible scheme of evolution of central nervous system, Plate IV
   neurotransmission in, 161, 166, 180
   nicotine reactions, 175–8, 207ch9n23
   and pharmacology, 137, 153, 158, 159, 164, 175–95
      author's research using flatworms, 181–3
      beginnings of research on, 178–83
      exhibiting withdrawal-like behaviors, 177–8, *178*
      numbers of research papers on planarians, *181*
      testing parthenolide against cocaine in, 190–2
   physiology of, 159, 160, 163, 164, 168, 175–6
   planarian neurons and neurons of vertebrate origin, 160–1
   planarians reacting to cocaine, 178, *179*, 207ch9n23
   and poetry, 137–8

in popular culture, 137–48, *146*, *148*
  on cover of *Scientific American*, Plate IX
  Planarian Man (superhero), 139–41, *142*
as predators, 106–7, 111, 158
  cannibalistic nature of, 107, 140
  as first hunting organism, 158
reaction to stimuli, 105, 172
  light reaction/perception, 105, 139, 170, 206ch8n2
  tactile stimuli, 169–70
and regeneration, 128–32, 140, 145
  ability to regain power after X-rays, 132
  of the brain and nervous system, 165–8
  head regeneration, 128, 204ch7n18
  impact of regeneration on ability to learn, 174
  impact on ability to learn, 172–3
  planarians and genomes, 132–6
  record number of pieces regenerating, 128–30, 205ch7n23
  story of the many cycles of regeneration, 197
regeneration, number of cells needed to allow for regeneration, 130, 205ch7nn23,24
regeneration in, 98–9
representative morphology of, 106
similarity to vertebrates, xiii, xiv, xxiv
in a state of "constitutive adults," 136
transplants in, 98–9
transporting of, 135, 205ch7n31
as triclads, 97
usage of the term, 102–3
use of in research, xi, 121–5, 176, 180
  applying planarian research to vertebrates, 192–4
  beginnings of research on planarians in pharmacology, 178–83
  choice of planarians as animal model in neuroscience, 164–8
  numbers of research papers on planarians, *181*
  reasons for using planarians in, 176, 180
  withdrawal-like behaviors in planarians, 177–8, 182
Planarians tasmaniana, 212n1
planariologists
  early works of, 108–11

  modern works on (author's personal connections to), 111–17
  usage of the term, 111–12
planning and predation, 156
"Plant Neurobiology, Intelligent Plants or Stupid Studies" (Rehm and Gradmann), 61
plants, 25, 59–61, 83, 88, 155
  and behavior, 21, 58–9, 60, 61, 127
  carnivorous plants, 59, 155
  combatting predation by herbivores, 156
  and defensive compounds, 35
  distinguishing animals from plants, 127
  intelligence in, 60
  neurotransmission in, 59
  not having neurons, 59, 60
  and pharmacology, 82–9
  plant electrophysiology, 59–61, 155
  reasons why plants create psychoactive substances, 82–3, 85, 88–9
  and regeneration, 126, 127
  that produce cembranoid-like molecules, 188
plasticity of the brain, 64, 125
Plato, 69
Platyhelminthes, 96, 99, 158, 160
pluripotent cells, 132
PNS. *See* peripheral nervous system (PNS)
poetry and planarians, 137–8
poison. *See* toxins; venoms
polarization of cells, 52
Polycelis felina, Plate VII
Polycelis sapporo, Plate I
polyclads, 96, 97–8, *Plate V*
polymers, 21, 22
polypetide, 23, *23*
pons, 62
popular culture, xxiv, 137–48
postsynaptic side of a synapse, 54, *55*, 75
  hyperactivation of dopamine autoreceptors, 87
potassium, 51
  sodium-potassium ATPase, 51–2
Precambrian period, 102
predators. *See also* carnivores
  ability to survive against predation, 16, 21, 35, 125, 154, 156, 184, 186
  cnidarians as, 126–7
  cone snails as, 36

predators (Cont.)
  planarians as, 111, 158
  plants as, 59
  social hunters vs. solitary hunters, 156
  Turbellarians as, 97
predictions, based on scientific Theory, 13–14
presynaptic side of a synapse, 54, 56
Prialt. *See* Ziconotide (Prialt)
primary metabolites, 84
problem-solving skills, 152
Proboscidea, 110
proboscis. *See* pharynx
procaine, 190, 191, 193
Proceedings of the Brünn Society for the Study of Natural Science (journal), 120
Prodromus (Müller), 103, 105, 108
prokaryotes, 20, 24–5, 152. *See also* cell biology
proliferation, 123
"Proposal of a New Model with Dopaminergic-Cholinergic Interactions for Neuropharmacological Investigations (Palladini and Carolei), 179
proseriata, 96
prostaglandins, 188–9
proteins, 19, 20, 22, 29–30, 35, 56, 74, 82, 83, 91, 165. *See also* macromolecules; neurotransmission; peptides (polypeptides); polymers; receptors
  and abused drugs, 181, 182
  amino acids in, 22–3
  ATPase, 51–2
  FGFR-related gene, 165
  and gap junctions, 50
  and ion channels, 53, 80, 185
  nictonic acetylcholine receptor, 185–6, 211ch10n39
  not alive, 29–30
  receptors, 20, 55–6, 80
  voltage-gated ion channels, 53
Protists, 25, 106
protopsychology, 168–74
Prozac, 57
psychohistory, 143
psychology, xxiv, 41, 87, 169, 170
  behavioral neuroscience, 170–1
  protopsychology, 168–74

psychopharmacology, 82–9
pure science, 34

QS. *See* quorum sensing (QS)
quantitative inheritance, 119
quorum sensing (QS), 151–2

radial symmetry, 99–100, 100
radiator, belief that brain acted as, 69
radioligand binding, 211ch10n39
Raffa, Robert B. (Bób), xi–xii, 180–1, 182
Ramón y Cajal, Santiago, 46–50, 121, 200ch3nn2,4
Randolph, Harriet, 121, 132, 134
rats, use of parthenolide against cocaine in, 192
reality, and the flavor of coffee, 8
receptors, 55–6, 74, 75, 79–82, 184, 191, 194. *See also* binding; communication; ligands; signaling
  autoreceptors, 74–5, 81, 87
  binding of transmitters to the receptor, 56
  FGFR-related gene, 165
  interaction between compound and molecular target, 79–81
  ligand-receptor interactions, 80, 81–2
  nicotinic acetylcholine receptor (nAChR), 85, 185, 186, 189, 190, 191, 211ch10n39
  postsynaptic receptors, 54, 55, 75, 87
  and proteins, 20, 55–6, 80
  receptor targets, 184
  serotonin type II receptor, 190
Reddien, Peter, 132, 135, 136
reductionism, 6–12, 46
  examples of
    flavor of coffee, 8
    learning about cars, 10–11
    wetness, 7–8, 10
regeneration, xiii–xiv, xxiv, 98, 111, 121–5, 123–32, 134, 164, 204ch6n34, *Plate VII*
  Darwin conducting regeneration experiments, 212n1
  neoblast theory, 131
  pathways of regeneration, 123
  of planarians, 128–32, 140, 145
  of the brain and nervous system, 165–8
    head regeneration, 128, 204ch7n18
    impact on ability to learn, 172–3
    planarians and genomes, 132–6

record number of pieces regenerating, 128–30, 205ch7n23
use of by different organisms, 125–6, 204ch7n18
"relaxing" effect of nicotine, 177
religion, 44, 45
reproducible results, 34, 169, 177
reproduction, 17, 23, 83, 120, 129, 152
as basis for survival, 16, 21, 42, 90, 168, 198
as criterion for choosing research animals, 90, 122, 134
reprogramming of cells, 132
research. *See* biology, biological research; biomedical research
resting potential, 52
Resurrection Ship (fictional), 144
reticularist theory, 46–50. *See also* Golgi, Camillo
"rewind and replay," 133
rhabdocoels, 96
rheotaxis, 106
ribonuclease, 173
ribonucleic acid (RNA), 21–2, 23, 25, 173
ribose, 21
Rimacephalus arecepta, 98, 202ch6n2
RNA. *See* ribonucleic acid (RNA)
Roadrunner and use of scientific names, 25–6
Romankenkius, 116
Romero, Rafael, 166
roundworms, 29
Rutherford, Ernest, 9

Saccharomyces cerevisiae, 29
Sagan, Carl, 182
Sagredo (fictional character), 209ch9n45
salamanders, 155
Saló, Emili, 205ch7n23
Salviati (fictional character), 209ch9n45
Sánchez Alvarado, Alejandro, 133–5, 136, 166, 205ch7n31
San Diego Reader (newspaper), 140
Sarnat, Harvey B., xiii–xv, 158
Sauerman (fictional character), 171–2
"scavenger" proteins, 56
Schickley, Timothy, 180
Schmidtea mediterranea, 133, 134–5, 136, 175, Plate VII
The Scholar's Journal, 110

science, 3–18. *See also* natural science; specific types, i.e., biology, neuroscience, etc.,
and the accepting of new ideas, 50
existence of, 3–6
honesty in reporting science, 108–9
as a human construct, 6
process of scientific discovery, 44–5, 183
reductionism, 6–12
as self-correcting, 33–4
science fiction and planarians, 137
Battlestar Galactica (TV show), 141–5
Dr. Who (TV show), 147
Fringe (TV show), 145–6
Planarian Man (superhero), 139–41
Scientific American (journal), 169
cover showing a planarian, Plate IX
sea anemone, 153–4
sea slug, 155
secondary metabolites, 84
The Secret Life of Plants (Tompkins and Bird), 60
segmental organization of the brain, 62
Seidlia auriculata, Plate I
Seldon, Hari (fictional character), 143
Selye, Hans, 19
senescence (aging), 136
sensory receptors, 157
serotonin, 54, 55, 59
serotonin transporter (SERT), 87
serotonin type II receptor, 190
sexual reproduction. *See* reproduction, 120
Shanidar Cave, 89
Sherrington, Charles, 53
sickle cell anemia, 91
sight of planarians. *See* eyes
Sigino, Hisao, 115
signaling, 55, 58, 59–60, 76, 80, 152, 153. *See also* communication; receptors
Silber, R. H., 145
siliceous shale, flatworm fossil in, *103*
Simarro, Luis, 47–8
simple systems, 7, 7
Simplicio (fictional character), 209ch9n45
slime molds, 152
slug, 152
small molecules, 183, 184
"Smeds." *See Schmidtea mediterranea*
Smith, Stephen, 201ch4n11

Smithsonian Institution, National Museum of Natural History, 116
sodium, 51, 56, 185, 186, 191
  sodium-potassium ATPase, 51–2
"solar-powered slug," 155
Spanish names, 200ch3n4
speciation, 120
species (as category in classification), 25
Spicilegia Zoologica (Pallas), 108, 109
spiders, 27, 35, 156
spinal cord, 62, 64, 65, 95, 159, 161
  injuries to, 125, 131, 165
sponges, 125, 128, 153
Sporns, Olaf, 72
squids, 29, 90, 151, 168
squirming, 178
starfish, 99–100
statins, 34–5
stem cells, 129, 132
Stevenson, Robert Louis, 140
Stower Institute, 133
"The Strange Case of Dr. Jekyll and Mr. Hyde" (Stevenson), 140
structural biology, 11
strychnine, 153
subjectivity and perception, 8
subsynaptic connections and interconnections, 74
sulci, 66
superhero, planarian as, 139–41
survival of the fittest, 16–17, 21, 198. *See also* evolutionary biology
symbiosis, 15, 127, 151, 153–4, 155, 184
symmetry, 99, 99–100, 100, 100. *See also* bilaterian animals
synapses, 42, 43, 53–7, 55, 73
  chemical synapses, 53–7, 54, 55, 74–5, 87, 153
  drugs that affect, 57
  electrical synapses, 53, 153
  extrasynaptic dopamine autoreceptor, 87
  numbers of in brain, 72, 76, 202ch4n16
  postsynaptic receptors, 54, 55, 75, 87
  presynaptic neurons, 54, 56, 57
  subsynaptic connections and interconnections, 74
  synaptic cleft (synaptic gap), 54, 55, 56, 87
  synaptic mapping, 73, 74, 76
  synaptic vesicles, 54, 55, 57, 161

syndesm as original word for synapses, 53
systematics, 24

T. *See* thymine (T)
tactile stimuli. *See* touch, sensitivity to
tailtwist, 178
Tanacetum parthenium, 190
Tasmanoplana tasmaniana, 212n1
taxonomy, 9, 24, 26, 95, 99. *See also* classification system for life
tectum, 62
tegmentum, 62
television shows and planarians
  Battlestar Galactica, 141–5
  The Big Bang Theory, 147
  Dr. Who, 147
  Fringe, 145–6
Temple University, 180–1
termites, 188
thalamus, 62
thalidomide, 32
theory (as a guess) vs. Theory (as a logical model), 12–14
Thompson, Richard, 171, 172, 208ch9n44
thymine (T), 22
Time Lords (fictional characters), 147
titin (protein), 23
tobacco cembranoids, 188, 192
Tompkins, P., 60
Torpedo californica, 186
touch, sensitivity to, 105, 169–70
toxicology, 79, 181
  as pharmacology's sister science, 180
toxins, 35, 57, 154, 156
  of the cnidarians, 126–7
  of the *Dilepti*, 106
  lophotoxin, 184–7, 185
  many polyclads are poisonous, 98
transcription, 22
translation, 22
transverse commissure, Plate VII
Trematoda, 96, 96
Trembley, Abraham, 126–7
triclads, 96, 97, 97, 98, 102, 104, Plate V
Triops, 107
"truly alive" experiments, 28
Tschermak, Erich von, 120
Tufts University, 174
turbellarians, 97, 98, 102–3, 103, 104, 106, 116

macroturbellarians, 97
microturbellarian, 97, 130–1, Plate III
Twain, Mark, 26
Twilight (movie and book series), 146
The Twilight Zone (TV show), 145

U. *See* uracil (U)
ultrastructure, 11
Understanding Human Behavior (McConnell), 172
Universidad Central del Caribe, 189
Universität Graz, 115
University of Barcelona, 135, 165, 166
University of Chicago, 113
University of Illinois, 133
University of Ljublijana, 115
University of Michigan, 171
University of Nebraska, 139
University of Puerto Rico (UPR), 115, 192
University of Rome, 179
University of Texas, 171
University of Virginia, 115
UPR. *See* University of Puerto Rico (UPR)
uracil (U), 22
urbilaterian ancestor, 101, 130–1
urea, synthesizing, 86
urochordates, xiv

Vanessa (fictional girl finding a planarian), 197
van Oye, Paul, 170
variable genetic populations, 30, 31–3
velvet worm, 95
venoms, 27, 35–6, 37, 156
ventral nerve cord, 161–2, Plate VII
Venturini, Giorgio, 179
Venus flytrap, 59, 155
Vermes, 95–6
Verrall, Arthur W., 53
vertebrates, 28, 29, 59, 85, 90, 95, 132, 133–4, 186. *See also* animals; invertebrates; specific names of species, i.e., gorillas, zebrafish, etc.,
  brains of, 62, 158, 160, 161, 192
  last common ancestor for (urbilaterian), 101

nervous systems of, 63, 64, 88, 125, 128, 159, 160, 164, 168, 177, 179
evolution of, xiv, Plate IV
photosynthesizing vertebrates, 155
and planarians, 156, 192–4
similarity to vertebrates, xiii, xiv, xxiv, 160–1, 162, 164, 166
use of in research, 192–5
vertebrate pharmacology, 153, 164, 176, 180
vinegar and oil, 51
viruses, 23, 25
voltage-gated ion channels, 53
Vorticeros praedatorium, 106

Waldeyer-Hartz, Wilhelm von, 42, 48
water fleas, 107
water molecules getting wet, 8, 199ch1n5
water perturbation, impact of on planarians, 169–70
Watson, James D., 9
wetness, 8, 199ch1n5
What a Plant Knows (Chamovitz), 61
"Where is Yesterday?" (McConnell), 171–2
Whewell, William, 4
Whöler, Friedrich, 86
Wiley E. Coyote and use of scientific names, 25–6
Willstätter, Richard M., 87
Wilson, Edward O., 9, 35, 209ch10n1
withdrawal-like behaviors in planarians, 177–8, 178, 180, 182, 192
Wolff, Etienne, 131
Worm Runners Digest (journal), 174
worms, generic, 95

X-files (TV show), 145

yeast, 29

zebrafish (*Danio rerio*), 28–9
Ziconotide (Prialt), 36–7
Zoological Museum of Amsterdam, 115
zoopharmacognosy, 78–9
zooxanthelae, 184
zygote, 129, 132